Like Ants

Like Ants

A Model for Human Civilization?

Andy Turnbull

Algora Publishing
New York

Library of Congress Cataloging-in-Publication Data —

Names: Turnbull, Andy, author.
Title: Like ants: A model for human civilization?/ Andy Turnbull.
Description: New York: Algora Publishing, 2017. | Includes bibliographical references.
Identifiers: LCCN 2017037754 (print) | LCCN 2017039152 (ebook) | ISBN 9781628943085 (pdf) | ISBN 9781628943061 (soft cover: alk. paper) | ISBN 9781628943078 (hard cover: alk. paper)
Subjects: LCSH: Human evolution—Social aspects. | Behavior evolution.
Classification: LCC GN281.4 (ebook) | LCC GN281.4 .T88 2017 (print) | DDC 599.93/8—dc23
LC record available at https://lccn.loc.gov/2017037754

Printed in the United States

Table of Contents

Top Dogs

Most people think humans are the dominant species on Earth, but that's a conceit. We are the most destructive species but we are not the oldest, the largest or the one most likely to survive after we have destroyed most of nature.

There are about seven and a half billion of us, and we have been around, as humans, for perhaps a million years. Biologists tell us there are about ten thousand trillion ants in the world — so many that their total weight is about the same as all of humanity — and they have been around for more than 100 million years.[1]

Among the 9,500 species of ants that have been identified, some hunt other insects, some scavenge, some enslave other varieties of ants, some tend herds of aphids and some grow fungus in underground farms. Most live in carefully engineered nests, many with their own water supply and engineered ventilation. Some nests are underground, some are tunneled into living or dead trees, some are made of leaves sewn together to form an enclosed hammock in living trees and some are made of forest litter. Some ants never come up to the Earth's surface and some never come down to the ground. Most have a permanent home nest but some are nomadic — spending most of their time on the move and stopping only to bivouac for a few weeks while their queen lays eggs.

Ants watched the rise of the dinosaurs and saw them die out. They saw the rise of humanity and they will probably be here to wave goodbye when we die out.

Biologist E. O. Wilson says the leaf cutter ants of Central America have perhaps the most complex insect society on Earth, often with millions of insects living in an underground nest that may be as big as a school bus. We see them as lines of workers, each carrying a section

[1] Hölldobler, Bert and Edward O Wilson, *Journey to the Ants*, The Belknap Press of HUP, Cambridge, Mass, 1994, pg. 1.

of leaf, and scientists tell us that a single colony consumes about as much vegetation as an adult cow.

They drink sap but they don't eat leaves. Instead, they carry them back to the nest where a complex production line converts them into an edible fungus.

The production line starts when scouts choose a tree to cut up but, before the harvesters start work, smaller inspectors check out the leaves for harmful bacteria and/or parasites.

If the inspectors approve the crop harvesters cut sections of leaves but, as they carry their burdens back to the nest, they are harassed by parasitic Phorid flies that try to deposit eggs that would hatch into flesh-eating maggots. Small guardian ants ride on their big sisters or on the leaf sections to protect the harvesters.

In the nest the sections of leaf are cut into smaller pieces by the largest class of gardener ants, then smaller gardeners chew the sections into a mass which they fertilize with their own waste, still smaller ones use the chewed mass to build a substrate for a garden and the smallest ones tend the gardens. A typical nest might have 100 or more gardens, each containing a ball of fungus about the size of a head of lettuce and weighing up to a kilogram. This particular fungus is found only in the nests of leaf cutter ants.

Other castes of ants build and maintain the nest, still others take out the waste and squads of soldiers, with mandibles that can cut leather, stand ready to repel invaders.[2]

Many colonies of ants war on others but some cooperate — joining nest to nest in super-colonies that dwarf human cities in population and even area. One 675-acre super-colony of 45 nests in Japan is estimated to contain 306 million workers and about a million queens, and the biggest single super-colony discovered to date stretches about 6,000 kilometers around the coasts of Spain and Portugal and across the French Riviera to Italy.[3] I'll leave it to someone else to count the residents.

We might find it hard to think of a series of nests covering several hundred acres, let alone 6,000 km, as a single colony, but myrmecologists (biologists who study ants) have a simple test. If an ant enters another colony's nest, it is usually killed; but any ant from any part of a super-colony will be accepted in any other part of the same super-colony.

The second most successful species on Earth is the termites. They look and behave so much like ants that many people think they are a species of ant, but in fact, they are quite different. Both are insects but while ants are descended from wasps and mostly carnivorous, termites eat cellulose and are descended from cockroaches. You can

[2] Wilson, E. O., *The Meaning of Human Existence*, Liveright Publishing Corp, WW Norton, NY, 2014, 98.

[3] "Ant super-colony dominates Europe," *BBC News*, Tuesday, 16 April, 2002. http://news.bbc.co.uk/2/hi/science/nature/1932509.stm

tell them apart if you remember that ants have a 'wasp waist,' but termites do not.

While ants raise males only to mate with virgin queens and do not expect them to work in their short lives before they leave the colony to perform their function, male termites that do not mate with a queen become asexual workers, called 'pseudergates,' which are more or less interchangeable with female workers.

Aside from the fact that they are both insects, the one thing ants and termites have in common is that both are what scientists call 'eusocial.'

That's a word University of Kansas entomologist Suzanne Batra coined in 1966 to describe the life style of the halicitine bees she was studying in India. Honey bees and bumble bees are eusocial too, but most of the 4,000 or so species of bees that have been identified live one or two to a nest and don't make honey.

Halicitine bees are small black 'sweat bees' that seem to love human sweat and that live in colonies like ants or honey bees. Batra's term was adopted and refined by Harvard University entomologist E.O. Wilson, who used it extensively in his classic book *The Insect Societies.*[4]

Biologists now use the term to describe animals living at the highest level of socialization, with two or more generations in nests or colonies that they will defend against intruders and that will not be abandoned when the first occupants die; with a caste system, division of labor and only one or only a few breeders. In most eusocial colonies, only the queens breed, and the young are cared for by workers rather than by their mothers.

There's no question that eusociality is the winning evolutionary strategy. Wilson tells us that the 20,000 or so known species of eusocial insects — mostly different varieties of ants, termites and some bees and wasps — make up about two percent of all known species of insects, but about three quarters of all insect biomass.[5] Between them, ants and termites make up about half the world's insect biomass.[6]

Most biologists used to accept (and many still do) the kin selection theory which assumes that eusociality develops because of a kin relationship among colony members. That seems to be the case among social bees and most varieties of ants, in which the workers of a colony are more closely related to each other than they are to their mother.[7] They all have the genes they get from their mother but they

[4] Wilson, E O, *The Insect Societies*, Belknap Press of Harvard University Press, Cambridge. Massachusetts, 1971.
[5] Wilson, E O, *The Meaning of Human Existence*, Liveright Publishing Corp, WW Norton, NY, 2014, pg. 47.
[6] Ibid., pg. 20.
[7] http://www.bumblebee.org

also have the genes they get from their father — which their mother does not share.

If a colony of eusocial insects has only one queen and she has mated only once, the workers of the colony will have all of their genes in common with each other but only two thirds in common with their mother. That sounds weird, but there is an explanation.

Female ants and bees hatch from fertilized eggs but males hatch from eggs that have not been fertilized, so the males have only half as many chromosomes as females. While most animals get half their genes from a father, worker bees and ants get two thirds of their genes from their mother and only one third from their father. Because the queen's father is not the father of the workers — who all have the same father and mother — the workers' relationship to each other is closer than their relationship to their mother. If the queen has mated more than once some have a different father, but their relationship to each other is still at least as close as their relationship to their mother.

The members of most colonies of ants and bees are more closely related than other families but, unfortunately for the kin selection theory, this does not apply in every eusocial species or even to every colony of ants or termites.

Wilson, for example, notes that termites will sometimes accept un-related termites into their colonies[8] and that some species of ants enslave, and keep and protect, other species of ants.[9]

Most of the slaves are captured as larvae and raised as members of the colony they serve. Still, there is some evidence that ants enslaved by other species of ants may rebel, and neglect or kill the young of their mistresses' colony.[10]

Fire ants raid each other's colonies and carry off the eggs and larvae. Biologist Walter Tschinkel of Florida State University tells us that if a series of raids are successful, the adults and even the queen of the nest that is robbed will join the robbers in their nest.[11] Eventually the two queens will compete and the workers will kill one; but the kin selection theory fails among fire ants and, obviously, among the ants that form super-colonies with multiple queens.

Wilson suggests that eusociality begins when a species builds a permanent defensible nest where it will raise its young and from which it can hunt and forage for food. Much as I admire Wilson I have to question him on this point because the army ants of Central America and at least one species of aphid-herding ants in Indonesia

8 Wilson, E O, *The Meaning of Human Existence*, Liveright Publishing Corp, WW Norton, NY, 2014, pg. 72.

9 Ibid. pg. 88.

10 http://www.world-science.net/othernews/120926_ants.htm

11 National Geographic film *City of Ants*, written by James Manfull and produced by him with Martin Dohrn and Stephen Dunleavy. https://www.youtube.com/watch?v55tXhnlZoOg

have no permanent nests. Both are nomadic and while they may stop for a few weeks while the queen lays her eggs, they are usually on the move.

The Herd

Eusociality seems to be the way to go but, Wilson says, only about twenty of the millions of species that have evolved on Earth have made it to eusocial status. They're easy to count because animals that don't become eusocial die out, sooner or later, but eusocial animals survive. Termites have been around for about 150 million years and ants for about 100 million, and ants evolved from wasps that may have been, like their modern descendants, eusocial.

Wilson estimates that the average life span of a mammalian species is about 500,000 years[12] so it's no surprise that we don't see many wooly mammoths, saber-tooth tigers or giant sloths in the modern world. Man has done better — a million years or more in our present form — so we may be doing something right.

Biologists used to think that only ants, termites, wasps and some species of bees were eusocial but, in the past few years, other eusocial species have been recognized. They include an Australian ambrosia beetle that lives in tunnels excavated in trees, some types of aphids and thrips that live in galls on trees, several species of spiders that share webs (including one species in Madagascar that numbers up to 50,000 spiders in a single network of webs) and three species of parasitic shrimp that live in nests excavated in sponges and even a few mammals.

The first mammals to be recognized as eusocial, naked mole rats live in clusters of 75 to 80 individuals, in the tropical grasslands of southern Ethiopia, Kenya, and Somalia. Their complex desert burrows may include three to five kilometers of tunnels.

The colony is ruled by a queen — the only fertile female — who may have up to three male consorts and who is supported by workers that gather food, build and maintain the nest, care for her pups and mount a defense if the nest is attacked. Worker males and females appear to do similar jobs.

Female workers are sterile, with immature ovaries and lacking some hormones, but male workers appear to be potentially functional. Queens live from 13 to 18 years and are hostile to other females that appear to be producing hormones. In experiments where the queen was removed or died, one of the non-reproducing females became sexually active, sometimes after a violent struggle with other candidates.

Wild queens usually breed once a year, producing a litter of 11 to 28 pups that weigh about two grams each, after about 70 days' gestation.

[12] Wilson, Edward O., *The Social Conquest of Earth* Liveright Publishing Corp, NY, 2012, pg. 14.

In captivity, they breed all year long and can produce a litter every 80 days. The young are born blind and the queen nurses them for the first month; then other members of the colony feed them feces until they are old enough to eat solid food.

The workers are divided into castes, with some specialized as tunnelers and some as soldiers. They eat roots and underground tubers, which they tunnel into and eat from the inside. Because the biggest tubers may weigh a thousand times as much as an individual rat and because they may keep growing while the rats eat them, a single tuber may last a colony for years. Sometimes the rats will block off a tunnel leading to a tuber, to give it a chance to regenerate.[13]

The Damaraland mole-rat of South Africa is physically larger than the naked mole rat, but otherwise similar. Both species live underground without ever coming to the surface[14] Mole rats are mammals but they are not warm-blooded. Most of us would describe them as cold-blooded, but, in scientific terminology, they are 'thermoconformers' — like snakes, toads and other animals whose body temperature is determined by the environment rather than by internal controls. They are also unique in that they have no pain receptors in their skins, and they use very little oxygen.

Meerkats, on the other hand, are unquestionably mammals. While not good house pets, they are cousins of the Indian mongoose which may, like *Rikki-Tikki-Tavi* in Kipling's timeless short story, be kept as an honored guest for its ability to kill venomous snakes.

Native to desert areas of Botswana, Namibia, Angola and South Africa, meerkats are small carnivores that look somewhat like small, starved, short-haired cats. They normally walk on four legs but you may have seen a picture of one standing upright, balancing on its tail and looking around.

They live in groups — called 'mobs,' 'gangs' or 'clans' — of 20 or more, up to about 50. For more than 20 years the Kalahari Meerkat Project, founded and headed by Cambridge University professor Tim Clutton-Brock, has studied several groups of meerkats and, the evidence shows, they must be considered to be eusocial.

A typical clan consists of one mating couple and their offspring, and the dominant couple will evict any females who become pregnant and/or kill their offspring. The young of the mating couple, on the other hand, are attended by full-time adult 'baby sitters' who may nurse them when they are very young and who teach them to hunt. Scorpions are among their favored food, and lessons include drill in the delicate art of biting the scorpion's deadly sting off before it gets a chance to use it. For the first few lessons with scorpions, the teacher bites the sting off first so the pup can practice with a scorpion that can't kill it.

[13] http://www.theanimalfiles.com/mammals/rodents/mole_rat_naked.html

[14] http://en.wikipedia.org/wiki/Naked_mole-rat. See also http://en.wikipedia.org/wiki/Damaraland_mole-rat

Because meerkats are prey as well as predators, they forage in groups, always guarded by sentries on the lookout for eagles and jackals. They seem to have a language of warning signals that distinguishes between flying and earthbound predators, whether the predator is close to the foragers or not and whether, and how fast, it is approaching.[15]

The fourth eusocial mammal is humans. That's not just my opinion — I take it from Wilson and many other biologists who study eusociality.

We defend our homes and in earlier times our towns and cities were surrounded by walls. Our countries have borders that are defended by armies and our houses have locks. Our towns and, through most of history, many of our houses have been occupied by three, four or more generations at any one time and we don't abandon a town or a house when the original inhabitants die. Even in a supposedly-democratic society we have castes and, for most of our history, we have had division of labor. Through most of our history parents have been primary caregivers to children, but often the job has been delegated to grandparents, neighbors and older siblings, and most of our nests — be they towns, feudal fiefs or nations — have recognized leaders.

How did we know what kind of behavior would be viable and what would not? The answer is that we did not know, and some early human lines died out — but some did not and we are descended from the ones that did not.

I suggest that all animal behavior can be sorted into two very general categories, which I describe as *survival positive* and *survival negative* behavior. Obviously there must be a very large gray area between extremes, but I ignore that for now.

Survival positive behavior is conducive to survival, and survival negative behavior is not. Consider, for example, the fate of two proto-humans who met a saber-toothed tiger.

Eek thought it looked dangerous, and he ran away. Ook thought it looked pretty, and he tried to pet it.

Eek had survival positive behavior, and he became an ancestor. Ook had survival negative behavior, and he was a snack. Because they were descended from Eek, most of our ancestors inherited his reluctance to pet saber tooth tigers.

Most examples of survival positive and survival negative behavior are less clear-cut than the case of Ook and Eek, but the principle holds. If you make mistakes you're out of the race, and every animal alive today is descended from ancestors that did not make fatal mistakes.

It's not as though Ook and Eek had different instincts. They probably had the same instincts but they made different choices, perhaps because they had different life experiences, and when Eek became a father he taught his children to react the way he had. We used to think that 'wild' animals are guided mostly by instinct, but

[15] kalahari meerkat project, http://www.kalahari-meerkats.com

modern zoologists know that while some behavior is instinctive; many animals must learn, from their parents or other elders, how to live.

Konrad Lorenz notes in *King Solomon's Ring* that one of his pet jackdaws, raised without elders to teach it, had no fear of cats and would actually try to perch on his dog's nose.[16]

In Panama, fringe-lipped bats (Trachops cirrhosis) eat frogs and toads; but some frogs are poisonous. The bats locate and identify their prey by listening to their calls and, in a study at the Smithsonian Tropical Research Institute on Barro Colorado Island in Panama, biologists Rachel Page and Mike Ryan found that young bats learned the calls of 'edible' frogs by noting which of the 28 species of frogs on the island other bats eat.[17]

The skin of a cane toad is poisonous but, in Australia, some rats have learned to eat only the toad's tongue and some varieties of ravens have learned to eat cane toads by reaching into their mouths, grabbing their tongues and literally turning the whole animal inside out. This can't be instinctive behavior because cane toads, native to Central and South America, have lived in Australia only since 1935, when they were introduced in an attempt to kill pests that attacked sugar cane.

Even grazing and browsing animals have to learn which plants are edible and which are not and, in some cases, how to get them. English biologist Kenneth Hall realized this when he watched a baboon that had been moved from a high-altitude forest in the Transvaal to a low-altitude game preserve on the Cape of Good Hope. Vegetation at the lower altitude was different and, until local baboons showed it which plants were edible and which were not, the newcomer did not eat.[18]

When domesticated cattle are released on a Western range they are sometimes poisoned by 'alkali' springs or by 'loco weed.' Bison and other 'native' animals seem to avoid these dangers but, if they do it by instinct, it is an instinct that non-native animals don't have.

Information about what to eat, what not to eat, and what to avoid being eaten by, might be termed the conventional wisdom of animals.

Galbraith coined the term in relation to human society, but because young animals learn from their elders — as young humans (sometimes) learn from their elders — we might assume that they have at least an analogue to conventional wisdom.

We associate the ability to learn with 'higher' animals but even fruit fly larvae can learn. In experiments, groups of larvae that had been given good food in an area that smelled of one chemical that was not related to food and bad food in an area that smelled of another

16 Lorenz, Konrad, *King Solomon's Ring*, Signet, New American Library edition, 1972, pg. 160.(first publication was by Thomas Y. Crowell, London, 1952.)

17 Page et al. "Social Transmission of Novel Foraging Behavior in Bats: Frog Calls and Their Referents." *Current Biology* 16, 1201 — 1205, June 20, 2006, DOI 10.1016/j.cub.2006.04.038.www.current-biology.com.

18 Cited by Ardrey, Robert, *The Social Contract*, Delta Books, (Dell), NY, 1970, p 82.

chemical — again not related to food — 64 percent of the test subjects would wriggle toward the smell that reminded them of good food.

In other experiments they were teased with a brush in an area with one smell and left in comfort in an area with a different smell. Given a choice, 73 percent would wriggle toward the area where they had not been teased.[19]

Many animals live in herds and we assume that this is instinctive, but I suggest that herding behavior might be learned. To me, it seems likely that animals that grow up in herds might prefer to live in herds. We might try to test this hypothesis by releasing some zoo-raised antelope on the African veldt to see if they form herds, but that might teach us more about the table manners of lions and leopards than about the herding instinct of antelope.

We might avoid that problem by setting zoo-raised antelope loose in a fenced preserve with zoo-raised lions and tigers, but that might not work either.

In a wonderful book, author Susan McCarthy describes the efforts of human care-givers on a reserve in South Africa to teach orphaned carnivores how to hunt. Tigers raised by humans were fed antelope but they would not kill one themselves until after their keepers hung a dead antelope from the back of a truck and made the tigers chase it and tear it loose from the truck before they could eat it. Their first real prey was a porcupine, but they didn't know how to attack it and, McCarthy says, one tiger might not have survived without treatment. I can imagine that because, as a news reporter, I once watched a 'coon-hound trial' in which a cast of hounds found and ate a porcupine. Vets who saved several of them said they found quills all the way down to the dogs' stomachs.

McCarthy writes that when some orphaned tigers first met a cow, the animal kicked them and drove them away. They were scared of ostriches but learned to hunt them after they were given a dead one. She reports that big cats raised by humans may never learn to hunt well enough to feed themselves.[20]

It seems that hunting is learned and, whether herding is instinctive or learned, animals in a herd gain more than company. Animals in a herd have dozens or hundreds of eyes to watch for predators and, when a predator does attack, only one member of the herd is likely to die. The same animal alone would be open to attack every time it stopped watching for long enough to eat, and when the attack came that one animal would die.

In times gone by, some antelope might not have joined herds, but only the ones that did survived to breed.

Herd animals can breed in such numbers that the few taken by carnivores are no threat to the survival of the species. As hunters

[19] McCarthy, Susan, *Becoming a Tiger, how baby animals learn to live in the wild*, HarperCollins, NY, 2004, pg. 6-7.

[20] Ibid. pg. 165, 175.

ourselves, we tend to admire carnivores, but biologists tell us that herd animals live longer and safer lives and that most of the herd animals taken by carnivores are either old or sick. When carnivores get old or are injured, they usually starve to death.

Birds gain an advantage when they flock, and fish when they swim in schools. Small birds can often out-maneuver big birds, so a raptor's best chance to catch a bird on the wing is to 'stoop' on it — diving so fast that the small bird has no chance to escape. It's safe to stoop on a bird that flies alone, but the raptor that stoops into a flock risks a collision that could break a wing. Even if the raptor takes a chance, it can only catch birds on the edge of the flock, and birds near the middle are safe.

Schooling fish have the same advantage. A small fish can out-maneuver and usually out-accelerate a big fish so, to be successful, a big fish has to attack with a rush — but the big fish that rushes into a school of small fish will collide with a ton or more of flesh and may be injured.

In an admittedly different context, many humans behave the same way. In an experiment, psychologists set up an artificial music market among more than 14,000 visitors to an internet web site. Visitors to the site were given a list of songs from little-known bands and asked to listen to any songs that interested them, and to download and rate the songs they chose.

About half of the participants were asked to make their decisions independently while the other half could see how many times each song had been downloaded by other visitors. Each member of the second group was randomly assigned to one or another of eight possible sub-groups which had evolved on their own; and the members of sub-groups could see only which songs other members of their sub-group had chosen.

In all eight sub-groups members downloaded songs that had been previously downloaded by other members of their group rather than songs that had not been as popular — but songs that did well or poorly in the control group, where people did not see other people's judgments, did very much better (or worse) than the same songs did in the sub-groups. The same song could be a hit if people downloaded it early in the experiment or a failure if they did not.[21]

Herding or flocking behavior may develop into cooperation and even self-sacrifice. The sentry that sounds an alarm calls attention to itself and it may be the predator's first target, but survival of an individual — especially of an individual male — is not important to a herd or to a eusocial community.

[21] Salganik, Matthew J., Peter Sheridan Dodds and Duncan J Watts, "Experimental Study of Inequality and Predictability in an Artificial Cultural Market," *Science*, 311, (2006)p 854-56. Cited by Thaler, Richard H and Cass R Sunstein in *Nudge, Improving Decisions about Health, Wealth and Happiness*, Yale University Press, New Haven, 2008, pg. 62.

Even suicidal behavior can be survival positive if it contributes to survival of a group or of offspring. I saw an excellent example of this in a nature film a few years ago. A pride of lions had cut a buffalo cow out of a herd and were about to kill it when a bull came to her rescue. The bull was a much more dangerous prey than the cow but the lions had to kill it, because when they attacked the cow the bull attacked them. When they attacked the bull, the cow ran back to the herd. Eventually the lions killed the bull, but it was a net gain for the herd.

The herd could afford to lose a bull because it had several, and the loss of one would make no difference to the next year's crop of calves. The loss of a cow would mean one fewer calf every year, and of the calves the cow-calves would bear.

I do not suggest that the bull was interested in the long-term welfare of the herd. I think it's more likely that he thought of the cow as property — or perhaps potential property — and he was protecting his property. His sacrifice served the herd because the cow was saved, so his action was an example of survival positive behavior for the herd, if not for the bull.

Wilson has been criticized for his contention that evolution can select groups as well as individuals but I think it obvious that while no individual survives for long, some groups do.

Some people call themselves 'social Darwinists' and think they have a natural right to prey on other people because the 'law of nature' is *survival of the fittest*. That sounds like a good excuse for predatory or uncooperative behavior, but it overlooks the fact that individuals don't survive. Groups and species do; and a group or species of animals that co-operate with each other have an obvious advantage over groups or species that prey on others of their own kind.

In fact we have solid evidence that the welfare of a group trumps survival of an individual in a classroom game that some sociologists used to use to demonstrate the tactical advantage of selfishness.

The game, called *Prisoners' Dilemma*, helped perpetuate the myth of the self-centered human for years — but the same game eventually disproved it.

The scenario of the game assumes that two criminals who have collaborated in several crimes are arrested and questioned separately. The police have enough evidence to convict both of a minor offense, but not enough to convict them of a more serious offense of which they are also guilty.

Each knows that if both keep silent, both will be convicted of the minor offense and each will be sentenced to one year in jail.

They also know that if one will confess to and testify about the more serious offense, he will go free and the other will be sentenced to three years in jail.

If both defect, both will be sentenced to two years in jail.

In the game each player chooses an option for each round. If both choose to keep the faith, they get two points each and if both defect they get only one point each. If one player defects and the other does not, the defector gets three points and the 'sucker' who keeps the faith gets one.

The number of points assigned to each choice may vary from game to game but the relationship is constant. A successful defector gets a high score, a cooperator gets a medium score and the victim of a defector gets a low score.

The assumption was that a good policy in the game is a good policy in life and that 'survival of the fittest' has given us instincts that naturally incline us toward the best policy.

A player who defects will always win against a player who does not and, in classroom sessions, most players looked for ways to fool their opponents into keeping faith while they defected as often as possible.

But Robert Axelrod, professor of political science at the University of Michigan, realized that short games could not tell the whole story. He argued that life takes more than a few hours to work out, and in the late 1970s, he invited games theorists to write their strategy for *Prisoners' Dilemma* into a computer program. Axelrod would then run the different games against each other in a tournament in which each game would consist of 200 moves — many more than students have time for.

Fourteen people submitted games and the tournament totaled 120,000 moves or 240,000 separate choices.

The winner was a program called *Tit for Tat*, written by Professor Anatol Rapoport of the University of Toronto. In the first round of each game *Tit for Tat* cooperates with its opponent and in each succeeding round it makes the same choice its opponent made for the previous move. It cooperates with any program that cooperates and defects on any program that defects.

Tit for Tat did not win a single game but it won the tournament on points because, even though it lost every game, it averaged a higher score than any other program.

In his analysis of the results Axelrod wrote of *nice* programs that cooperate when they can and *nasty* programs that try to win by defecting.

Eight of the 14 entrants in the tournament were nice programs and those eight took the first eight places. They won because they gained more points playing with each other than they lost against the nasty programs. The nasty programs won individual games but they lost the tournament because they lost more points when they played against each other than they won when they played against nice programs.

The results were so surprising that when Axelrod published them, he announced a new tournament. This time every one of the 62 entrants knew *Tit for Tat*'s strategy and that it was the one to beat.

It was no contest. *Tit for Tat* cleaned up and later simulations showed that would have won five of six possible variations of the tournament. In the sixth, it would have come second.

In later analysis Axelrod found a couple of variations of entries that would have won that particular tournament, but they were all nice programs and most of them would lose to *Tit for Tat* in most possible tournaments.

Animals win the evolutionary race by breeding more than others. In an 'evolutionary' version of the tournament, in which each program would be rewarded by having one extra copy of itself entered in competition for each point it earned, *Tit for Tat* would have crowded most other programs out.

Further analysis of the tournament results suggest that a small group of nice guys who move into an area dominated by nasty guys will prosper, but a small group of nasty guys who move into an area dominated by nice guys will not.

It would take a book to analyze all the implications of Axelrod's tournament (and Axelrod has written that book),[22] but we can sum it up in a few words, which I describe as 'Rapoport's law.' The evidence is that, in the long run, nice guys finish first.

After Axelrod's book was published Martin Nowak — then a grad student — and Prof. Karl Sigmund of the University of Vienna, developed a program that can beat *Tit for Tat;* but the new program, called *Generous Tit for Tat* is even nicer than the original. While *Tit for Tat* always retaliates against defection by always playing the same move as its opponent played in the previous round, *Generous Tit for Tat* sometimes forgives defections. [23]

Cooperation beats physical strength and even intelligence. Chimpanzees and gorillas are stronger and individually more intelligent than baboons, they are protected by law in many areas and they have no natural predators — but they are endangered and likely to become extinct. Baboons are not as strong or as smart as the apes; they are a favored prey of leopards and farmers shoot and poison them as pests — but they cooperate better than apes and they are not threatened.

South African naturalist Eugene Marais wrote that after a leopard kills or disables a baboon it runs and hides from the 'hit squad' of male baboons that will go after it. He also reports a specific case in which baboons attacked and killed a leopard that was stalking their band.[24]

[22] Axelrod, Robert, *The Evolution of Cooperation*, Basic Books Inc, New York, 1984.

[23] Nowak, Martin, with Roger Highfield, *Super Cooperators, Altruism, Evolution and Why We Need Each Other to Succeed*, Free Press (Simon & Schuster) NY, 2011, pg. 36+. Nowak notes that an essay by Robert Mays of Oxford University set him on the track to the new strategy.

[24] Marais, Eugene, *My Friends the Baboons*, pp 42-3. The book began as a series of articles in *Die Vaterland* newspaper, then a book published in Afrikaans as *Burgers van die Berge* by J.L. van Shaike, Pretoria, 1939, English translation by Blond and Briggs Ltd, London, 1975.

We don't think of humans as herd animals but we live in groups or bands, and it's not hard to understand why. We are one of relatively few animals that are both hunters and potentially prey for carnivores, and the formation of groups is survival positive for both roles.

We don't often think of humans as prey but a single unarmed human is an easy kill for any large carnivore. Still, in most of the world even a single unarmed human is safe from most carnivores because, like baboons, groups of men usually track down and kill any carnivore that kills a human. We've been doing that for thousands of years, and few carnivores that developed a taste for human flesh have survived long enough to teach their preference to others.

Herds don't have to organize teamwork but hunting packs do. Primatologist Jane Goodall described how chimpanzees hunt Colobus monkeys, which are smaller than chimps but more agile and much better climbers. A single chimp would have little chance of catching a monkey but, when a group hunts, members of the group casually climb trees surrounding the chosen prey before the killer climbs the tree the prey is in.[25]

If the killer were alone the prey could escape easily, by jumping or swinging to another tree but it is surrounded before it knows it is being hunted, and there is no escape. Without a plan, chimps could seldom, if ever, catch a monkey.

A single wolf could not kill a moose and in most conditions most deer could out-run most wolves but, hunting as a pack, some wolves live mostly on elk, caribou, moose or other large deer.

In winter wolves can run on crusted snow, while deer cannot. If there is no crust, deer can still outrun wolves but, if a pack of wolves follows a herd of deer, usually one deer will panic and, likely, run into a snowdrift or other trap.

In summer, wolves try to drive deer into dry or shallow riverbeds, where the wolves' pads work better than the deers' hooves and the deer may to stumble on round stones.

In a pack hunt, female wolves — which can run faster than the males — usually herd the prey and males — which are stronger — make the kill.

The alpha male may choose the prey and perhaps break off the hunt if it is not going well but, otherwise, each member of the pack chooses and acts its own role.[26]

Early humans had to cooperate to hunt because, without weapons, there are few animals an individual human could catch and kill. Even with the kind of weapons that were available to our ancestors, an individual has little chance of success but groups of men are very good hunters. Archaeologists think men with spears were able to kill mammoths, some Native North Americans drove whole herds

[25] Goodall, Jane, *In the Shadow of Man*, Collins, London, 1971, pg. 181.
[26] https://livingwithwolves.org/how-wolves-hunt

of buffalo over cliffs and the Ba'Aka Pygmies of the Central African Republic hunt duiker antelope by driving them into nets.[27]

Flocking, schooling, herding and pack hunting behavior are all survival positive, so it's no surprise that our ancient ancestors lived in herds and hunted in packs. Eusocial behavior is even more survival positive but while many species live in herds and hunt in packs, relatively few take the final step to full eusociality.

Even among herding and pack-hunting animals we had two big advantages over the others. One is that we are able to make more complex sounds than other animals and, with speech, are able to pass knowledge and advice from one individual to another more efficiently and probably more accurately than other animals.

Speech is a huge benefit because, as noted before, the key to eusociality is the division of labor, and speech makes it easier to divide labor. If you learn by imitation, you will learn only what you see someone doing. With speech, a person doing one job can tell you about a different job that needs doing.

Apes learn complex behavior from one another, but they learn by imitation. Anthropologist Carel Van Schaik reports that orangutans in Sumatra like and eat the nutritious seeds of neesia fruit but those seeds are protected by sharp spikes and hard to get at. Orangs on one side of the Alas River have learned to use sticks to dig the seeds out of the fruit but orangs on the other side of the river — which is too wide for them to cross — have not. Orangs on both sides of the river eat the seeds but the ones that have not learned to use sticks don't eat as many. Upstream, where the river is small enough for the apes to cross, it appears that orangs have not learned to eat the seeds.[28]

We also have the incomparable advantage of reading and writing, which enable us to learn from people we never met and even from people who died hundreds of years ago. This is not an un-mixed blessing, but it is a blessing.

An orang that knows the trick of eating neesia seeds can teach others only if he has a neesia fruit to demonstrate with. A human can explain the trick to other humans at any time, with no need to demonstrate it. If you were to go to Sumatra and find a neesia fruit, you know the trick, because you read it here. An orang from a North American zoo would not.

It's not crucial but it is also a huge advantage that we can control — and even make — fire.[29] Like other animals our ancestors must have seen fires caused by lightning and, like others, they probably

[27] http://travel.nationalgeographic.com/travel/traveler-magazine/unbound/africa-pygmies/.

[28] Von Schaik, Carel, "Why are some animals so smart?" *Scientific American*, special edition "Becoming Human," summer 2006.

[29] Biologist E.O. Wilson cites our two major advantages in this development as our size and the fact that we are land animals. He notes that ants are too small to manage fire and that dolphins, which may be as smart as some humans, have no hands and, living underwater, are obviously unable to manage fire.

scavenged animals and vegetables that had been cooked by fire — but our ancestors went a step further. For one reason or another — possibly because it was tastier — we learned to prefer cooked food and we had the ingenuity, and the nerve, to learn to control fire. The preference for cooked food is valid, because it's easier to digest and therefore more nutritious.

Fire is useful but not absolutely necessary. Steak tartare is a gourmet dish in some countries and I have eaten raw frozen caribou, raw walrus and 'muktuk' (whale blubber and skin) with Inuit friends, and raw fish — sashimi and sushi — in Japan and Canada. Once (but only once), I ate a raw lobster. Jerky, biltong, pemmican, salami and pepperoni are usually eaten raw.

Even after they learned to control fire, it must have taken our ancestors a long time — perhaps hundreds or even thousands of years — to learn to make it. They could take burning branches and glowing embers from natural fires back to a camp, and they could nurture and maintain camp fires, but it's a big step from there to learning how to make fire when and where you want it. It's easy to make a fire by striking sparks from flint, but you need steel to strike them with. Using a bow-drill to start a fire sounds simple to us, but we didn't have to invent it and we have modern tools to make it with. Inuit people who live north of the tree line and far north of any source of hardwood know how to make fire drills, but they have to find driftwood or trade with their southern neighbors for the wood to make them. Men have used fire for at least 400,000 years, but when Abel Tasman reached Tasmania in 1642, the people there could not make fire.

Because our very early ancestors could not make fire they often gathered in camps where a fire lit by natural means had been maintained. We might have become eusocial without the ability to control fire, but it seems likely that fire helped.

Not being able to make fire would have been a problem for early men because many of them were nomadic, following herds as they migrated and moving with the seasons as choice food plants ripened or fruited in different areas.

Being nomadic is not the same as being 'homeless.' Farmers and people who go to a job every day tend to have only one home but some kings and billionaires are nomadic, with houses or condos in several cities and a country estate, a tropical retreat, a summer home and perhaps a hunting and/or ski lodge. Early men may have moved from camp to camp, but they probably moved from one familiar campsite to another.

In my early teens I used to go on canoe trips with friends. We might be out for a few days or a week or more, but we didn't have to make a new camp every night. Instead, we stopped at familiar campsites with fire pits, prepared tent-sites and sometimes latrines. These were not commercial operations — they were on crown land and open to all — but they were left over from previous canoe trips; ours and others.

If one was occupied we could move on to the next campsite, but we seldom had to make a new camp.

When I travelled with a couple of RCMP 'Mounties' on a 'long patrol' in Canada's Northwest Territories nearly fifty years ago we spent a couple of nights in tents that had been left empty at campsites and, one night, we stopped at a campsite that was already occupied by a half-dozen or so traveling Gwitchen Indians. At one point we stopped to look (but not stay) at a crude shelter that someone had built when he was stuck and had to spend a night on the trail. Made of fir branches and plastered with snow, the shelter was crude and it would not have lasted long into spring, but many people knew of it and the Mounties had been told where to find it. The land that is wilderness to city people is more like a giant tourist resort to the people who live there.

Some campsites are more convenient than others. In my canoe tripping days, we often camped at portages — where we had to land and unload the canoes anyway, to carry them from one lake to another or to pass a waterfall. Traveling with the Mounties, I found that most campsites were at spots where the trail branched or where there was an obstacle to be negotiated.

Early hunter gatherers probably moved from one established camp to another and, in many cases, they could move into huts or other shelters they had made and used before. They could and probably did carry embers from one camp to another, but if they were too long on the trail, or if the embers went out, they would have no fire until they found another 'natural' fire or a campsite that had a fire.

A campsite with a fire would obviously be more desirable than one without but, with or without fire, our early ancestors probably had an easy life even when they lived in harsh conditions. In the 1960s, anthropologist Richard Lee studied the hunting and gathering !Kung San people of the Kalahari desert in the third year of a disastrous drought, while about 180,000 farmers of the Herero and Tswana tribes in the surrounding area had to depend on the United Nations' World Food Program for emergency relief.

Despite the drought and even though some Herero and Tswana gathered wild food in competition with them, the !Kung San lived in comfort and plenty, with very little work. Lee found that "a woman gathers in one day enough food to feed her family for three days and spends the rest of her time resting in camp, doing embroidery, visiting other camps or entertaining people from other camps. In a day spent at home, kitchen routines such as cooking, nut cracking, collecting firewood and fetching water occupy one to three hours of her time."

Men work longer hours than women, but their schedule is uneven. It is not unusual for a man to hunt for a week and then not hunt for two or three weeks.[30]

[30] Lee, Richard B. and Irven Devore, eds, *Man the Hunter*, Aldine Publishing Co., Chicago, 1968, pg. 37.

Most early hunter-gatherers were nomadic but anyone who did not want to move could probably have stayed in the same camp for months at a time, because both hunting and gathering would have been better near campsites than away from them. Even now, primitive villages and camps around the world are surrounded by clumps of fruit trees and berry bushes.

They grow because people who eat fruits and berries also eat some of the seeds, and if they do not walk away from the camp to defecate, they do to empty their chamber pots. People who do not eat the seeds throw them out, and their garbage midden becomes an orchard or a berry-patch or both.

Archaeologists in Southern Ontario, where I live, know that they can often find the sites of prehistoric villages or camps near groves of sumac trees — which produce berries that can be used to make a refreshing drink. It is generally considered more likely that the groves grew near camps than that camps were established near groves, but it could work both ways.

Modern Australian Aborigines are not farmers, but they are known to carry seeds of useful plants with them, and to plant them near favorite campsites.

When our hunting and gathering ancestors moved camp they probably moved from choice rather than necessity. Sir George Grey, one of the early explorers of Australia, noted that mimosa gum was a favorite snack food of the Aborigines and that some bands would travel a hundred miles or more to be on hand for the few weeks that a favorite tree exuded the gum.[31]

Some Mesolithic peoples used wild grains to make porridge and beer, but they did not have to farm them. Einkorn, a primitive form of wheat, grows wild in the Middle East and Mesolithic people could collect all they could use without farming. In the 1960s, University of Chicago Prof. Jack Harlan, working near the site of the early Turkish village of Cayonu, used a flint sickle to gather four pounds of wild Einkorn in an hour. His take threshed out to two pounds of clean wheat with 24 percent protein, compared with 14 percent for most modern wheat.

Based on this experience, he estimated that in pre-agricultural times an average family working for the three weeks of harvest could probably have collected about a ton of clean wheat.[32]

Prof. Harlan notes that virtually every known group of hunter gatherers knew enough about plants that they could have farmed at any time, but they had no need to.[33] If a single nuclear family could

[31] Sahlins, Marshall, *Stone Age Economics*, Aldine-Atherton, Chicago, 1972, pg. 7. See also Grey, Sir George. 1841. *Journals of Two Expeditions of Discovery in North-West and Western Australia, During the Years 1837, 38, and 39...* 2 vols. London: Boone.

[32] Braidwood, Robert J, 1975, *Prehistoric Men*, 8th edition, Scott Foreman and Company, Glenview, Ill. pp. 118-119.

[33] Harlan, Jack, 1995, *The Living Fields*, Cambridge University Press, pg. 15, 53 Random House, New York, 1977, pg. 36.

collect a ton of wheat for three weeks' work, it does not seem likely that anyone would voluntarily invest the months of work it would take to cultivate and protect a crop that might produce five or ten tons or even more. Better to gather one ton of grain and devote most of your time to hunting and collecting other foods.

The women of the band would probably prefer other foods because wheat has to be mashed or milled and then mixed with water before it is cooked into porridge or biscuits. Work for a woman who gathers roots and berries is light and pleasant but work for a wheat farmer's wife is hard and possibly painful. Archaeologist Elizabeth Barber says that toe, knee and shoulder bones of many women who lived in the early farming villages of Mesopotamia were deformed by kneeling and pushing as they ground grain with the type of stone grinders found in those villages.[34]

These deformities developed after people turned to agriculture and, I suggest, that came after other developments. Since physical anthropologists have learned to date human remains and to deduce the health of the living person from the state of the skeleton, they have found that most hunters and gatherers were healthier and lived longer than most farmers.[35]

That is partly because most hunters and gatherers have a better and more varied diet that than most farmers. People who grow grain eat grain but people who forage may eat hundreds of different plants, animals and eggs.

The Past

Most people who write about the future don't worry about the ancient past. That's history, they say, and we're looking at the future.

But most of our present and much of our future is a product of our past and, if we have misconceptions about the past, we can't understand the present or foresee the future.

Sailors know that a ship is not often going in the direction it appears to be heading. There are currents in the sea and if you want to head due east through a northerly current you have to head south of due east. Even steamships have to compensate for winds and currents, and sailing ships often have to 'tack' back and forth across the wind to get where the sailor wants to go.

So if you want to know where a ship is going, it's not enough to know where it is and which direction it is headed. You have to know where it started and where it is now, or where it is headed and which way the wind is blowing and how the currents are flowing, and how much the sailor knows about the wind and the currents.

[34] Barber, Elizabeth, 1984, *Women's Work -the First 20,000 Years*, W. W. Norton & Co, NY, pg. 96.

[35] Bezruchka, Stephen, "Is our society making you sick?" *Newsweek* Feb 26/01, pg. 14. See also Hayden, Brian, 1993, *Archaeology, The Science of Once and Future Things*, W H Freeman and Co. NY, pg. 220.

If a sailor sets his course without knowing about the wind and the currents, he won't get to wherever he wants to go. If pundits and politicians don't understand where we came from and the various factors that cause change, they don't know where we are going and can't plot a course to where we want to go. Remember George W. Bush's "Mission Accomplished" banner after his invasion of Iraq? He accomplished something all right but, to be charitable, we have to assume that he did not know what he had accomplished.

Historians study the past because, they say, 'history repeats itself' — so if the Persians invade Greece again, or Mongol hordes sweep out of Asia to conquer Europe, or if we discover a couple of new continents with lots of gold but no steel or gunpowder, they should be able to predict exactly what will happen.

But 'history' covers only a tiny fraction of our time on Earth. Depending on how you define humanity our species is anywhere from 100,000 to more than a million years old, but 'recorded history' covers only a few thousand years.

Archaeologists look at a longer span, but they too are bound by tradition.

The traditional view of human pre-history holds that early man came down from the trees to live on the African savanna and that our ancient ancestors were nomads, living on whatever they could find, until some bright soul invented agriculture. Then some of them settled in villages to support the farms and, finally, we developed civilization.

It was a good story, but it conflicts with modern scientific evidence and opinions. Most people don't question it because once a belief is enshrined in tradition, it's not easy to change. Even now, tens of millions of people believe that the world was 'created' in seven days and that humanity was 'designed' to rule it.

Belief

What we do or do not believe seems to be governed by five factors, that I call *primacy, authority, imitation, comfort* and *repetition.*[36]

Primacy refers to the fact that we generally believe what we hear first, and tend to ignore new information that conflicts with the belief we have accepted. This has been demonstrated by several experiments.

In one, subjects were told about a (fictitious) warehouse fire that had been caused by flammable materials improperly stored in a closet.

They were later told that the closet had actually been empty and, after the report and the correction, the subjects were given some 'distractor' tasks before they were asked some test questions.

[36] Prof Robert Cialdini of Arizona State University lists six factors in belief — *consistency, reciprocation, social proof, authority, liking* and *scarcity* in *Influence, The Psychology of Persuasion*, (Quill, William Morrow, NY, 1984, 1993, pg. xiii). I'm sure other people could name other factors, but I'll stick with my list here.

Even though they remembered the correction, some subjects suggested that the fire was hotter than normal because of the flammable materials, and others thought insurance coverage should be invalidated because of the stored materials.

In another experiment undergraduate psychology students were shown a fake police report about a minibus accident in which the bus was said to have been taking a group of elderly people back to their nursing home. This was later corrected with the information that the bus actually carried a group of college hockey players returning from a victory party.

Asked why it was difficult getting passengers out of the minibus, many of the students answered that it was because they were so old.[37]

This problem is more common among older people because, while older people may appear to have good memories, they may not remember where or how information was acquired.[38]

Primacy is a well-known phenomenon, and one that is sometimes used by sleazy lawyers who introduce false evidence in a court case and then correct it later. The false evidence will probably be remembered but, if he is caught, the lawyer can always show that he 'corrected' it.

It's also popular with propagandists. Many Americans still believe that Iraq was developing and had the 'weapons of mass destruction' that were cited as a pretext for the invasion of that country.

Advertisers can use this effect by publishing a false claim — in a news release or through a friendly reporter — and then correcting it. More people will remember the false claim than the correction.

The *authority* that presents information is also important. We tend to believe information that comes from sources that we recognize as authorities — such as established media, teachers and elders — and we discount information published in leaflets, on the internet or by people who have been labeled as 'rebels' or are not officially 'qualified.'

This is why actors in TV commercials pretend to be doctors, dentists or successful businessmen. The advertisers know that many people will accept them as authorities and, in a world dominated by media, actors, athletes and other 'celebrities' are also seen as authorities.

The power of *authority* is awesome, and ominous. In his book *Influence, The Psychology of Persuasion,* Professor Robert B. Cialdini cites an experiment in which a man claiming to be a doctor phoned a nursing station at a hospital and ordered the nurse on duty to give 20mg of the drug Astrogen to a patient.

---

[37] "Explicit warnings reduce but do not eliminate the continued influence of misinformation" Ullrich K. H. Ecker, Stephan Lewandowsky, and David T.W. Tang, *Memory & Cognition* 2010, 38 (8), pp 1087-1100.

[38] Glisky, E. L., Rubin, S. R., & Davidson, P. S. (2001). "Source memory in older adults: An encoding or retrieval problem?" *Journal of Experimental Psychology: Learning, Memory, and Cognition,* 27, 1131–1146. http://dx.doi.org/10.1037/0278-7393.27.5.1131

It was a direct violation of hospital policy for a doctor to prescribe a drug by phone; none of the nurses knew the 'doctor' who phoned or even whether he was a doctor or not; the drug Astrogen had not been cleared for use or placed on the ward stock list at that time and labels on the Astrogen containers specifically warned that the maximum dose was only 10mg/day — half the amount the supposed doctor ordered — but in 20 of the 22 hospitals in which the experiment was tried the nurse measured out the 20mg dose and headed for the patient's room where she expected to administer it. In each case an observer intercepted her and explained the experiment; so the potentially-fatal dose was never administered.[39]

Imitation is also important. As herd animals, we tend to believe what the herd believes and to model ourselves on public figures and even fictional characters.

In 1774 Johann Wolfgang Goethe published *Die Leiden des jungen Werther*[40] (The Sorrows of Young Werther) about a young man who wore boots, a blue coat and a yellow vest, sat at a desk with a copy of a popular romantic play, and shot himself with a pistol. In the next few years so many young men dressed as Werther and shot themselves in obvious imitation of him that the book was banned in several countries.

Psychiatrists call it the 'Werther effect' when one suicide in a mental hospital is followed by others and, forty years ago, American sociologist David Phillips found that the U.S. suicide rate — which at the time averaged 1,200 to 1,700 every month, depending on the time of year and other factors — increased by an average of nearly 60 in the month after any suicide reported on the front page of either the *New York Times* or the *Daily News*.

Some suicides have more impact than others. In 1962, the death of Marilyn Monroe apparently triggered nearly 200 extra suicides in the next month. In April of 1994 singer Kurt Cobain shot himself, and from then until the end of the year, quite a few teen-age suicides played Cobain tapes as they killed themselves and some left notes naming Cobain. Statistics also show that the frequency of single car motor accidents increases after a publicized single vehicle accident in which a famous person dies.[41]

Cialdini calls this 'social proof' and suggests that it is a factor in copy-cat murders and in the mass suicide of more than 900 people at the 'Rev.' Jim Jones' Jonestown colony in Guyana in 1978. The chosen

---

[39] Cialdini, Robert B., *Influence, The Psychology of Persuasion*, Quill, William Morrow, NY, 1984, 1993, pg. 224-5.

[40] Goethe, Johann Wolfgang, *Die Leiden des jungen Werther*, The Sorrows of Young Werther in the English translation by Harry Steinhauer, Norton, NY, 1970.

[41] Phillips, D.P. "The influence of suggestion on suicide, substantive and theoretical implication of the Werther effect." *American Sociological Review*, 39:340-354, 1974; see also Phillips, D.P. "Suicide, motor vehicle fatalities and the mass media," *American Journal of Sociology* vol. 84, no 5, pp. 1150-1174, 1974.

method there was poison, and many of the faithful saw their friends die in agony before they took the dose.

Our propensity to follow an example is well established. In the early 1950s, psychologist Solomon Asch conducted a series of experiments in which about three quarters of the subjects chose to agree with a group rather than to trust their own judgment.

Each run of each experiment involved one subject and six to eight others who pretended to be subjects but were in fact Asch's confederates. The experiment was described as a 'test of perception' and the group was shown two cards. One of the two had a single line drawn on it, the other had three lines. One of the three was the same length as the line in the single card, one was slightly but still noticeably different, and the third was very different.

Confederates and the subject were asked, in turn, to decide which of the three lines was the same length as the sample. In the first couple of runs the confederates gave the right answer but, in later runs, they began to give wrong answers. The subject of the experiment was always one of the last two or three to be asked and, after a couple of runs, about two thirds of the subjects agreed with the confederates, even on obviously incorrect answers.

Different runs of the experiment showed that if one of the confederates gave the right answer first, the subject was more likely to give the right answer.[42]

Other experiments since have confirmed Asch's results, and psychologist Irving Janis saw this tendency of decision-makers to agree with their leader as the cause of many bad decisions, including the lead-up to wars. In *Victims of Groupthink* he described it as "a powerful source of defective judgment that arises in cohesive groups, the concurrence-seeking tendency, which favors over optimism, lack of vigilance and sloganistic thinking about the weakness and immorality of out-groups... [T]he more esprit de corps in the group, the greater the danger of groupthink — which is likely to result in irrational dehumanizing actions directed against out-groups."[43]

Groupthink has also been cited as a probable factor in the loss of the space shuttle Columbia, and in the American invasions of Afghanistan and Iraq, which much of the world sees as unjustified aggression that sparked decades of war and terrorism.

But a leader does not act alone. The decisions attributed to groupthink are mostly the product of homogeneous and cohesive groups that work under considerable pressure. Janis argued that such groups may begin to consider themselves infallible and, if that happens, they will ignore the advice of experts and will tend toward extreme decisions. The driving forces in this are presumably a combination of *imitation* and, because we want to follow the herd, *comfort*.

[42] Asch, Solomon, "Opinions and social pressure," *Scientific American*, vol. 193 #5, Nov. 1955, pp. 31-35.

[43] Janis, Irving L., Victims of Groupthink, Houghton Mifflin & Co., Boston, 1972, p 13.

Comfort can also affect our religious beliefs. If one religion threatens damnation and another offers the prospect of heaven, many people will choose the heaven. As Francis Bacon once wrote, "Man prefers to believe what he prefers to be true."[44]

But the king of all factors is *repetition*. If an idea is hammered home by repeated presentations and especially if it is repeated in different media — newspapers, magazines, radio and TV programs and perhaps movies — it will probably be accepted on a subconscious level even if we think we reject it. This is the effect that advertisers count on when they present the same offensive advertisement time after time. You may hate the advertisement, but you are still likely to buy the product.

Advertisers and propagandists try to fool the public but I suggest that, because they live in a maelstrom of questionable information, they themselves may be more likely to believe their claims than we are. On the other hand, French scholar Jacques Ellul tells us that illiterate people living in remote areas are virtually immune to advertising and propaganda.[45]

Conventional Wisdom

Once beliefs are established, they become conventional wisdom, which can be defined as a set of beliefs that most of the members of a given society accept without question. It has been a guiding force in the development of all human cultures since the beginning — probably since before we were actually 'human.'

Among our early ancestors conventional wisdom might have included such gems as 'don't eat the mushrooms with the bumps on them; don't tease the big orange cat with the black stripes; don't swim near the big lizards with the bumpy backs' and 'don't stand on top of a mountain and shake your spear at a thunderstorm.'

That was good advice and people who did not eat toadstools, tease tigers, swim with crocodiles or shake their spears at thunderstorms lived longer and raised more children than those who did. Children who took this kind of advice were more likely to grow up and raise children of their own than those who did not, and we developed a tradition of listening to the elders of our band and ignoring most others. Most of the early conventional wisdom must have been useful but, through the ages, some conventional wisdom was created by leaders and priests for their own benefit. It worked for the people who created it but was not always beneficial to all.

The Inca used to drug children and leave them on mountain-tops to freeze to death as gifts to the gods. The Maya, another 'civilized' people, used to weigh teen-age girls down with gold jewelry before

[44] http://www.goodreads.com/author/quotes/50964.Francis_Bacon

[45] Ellul, Jacques, *Propaganda, the Formation of Men's Attitudes*, Alfred A Knopf, NY, 1965, trans by Konrad Kellen and Jean Lerner, pp 109-10.

throwing them into wells — again as gifts for the gods. As this is written, thousands of people are being slaughtered and millions displaced by a war which, on one level, is based on a dispute about which of two priests who lived more than a thousand years ago was the legitimate successor to the prophet Mohammed.

Two hundred years ago many people believed that witches cast spells and danced with the devil, and that both witches and heretics had to be burned alive to save their souls. A hundred years ago we knew that women and people with brown or black skins are less intelligent than white men.

Seventy five years ago most Americans knew that front wheel drive cars would never be practical and that the Japanese could produce only cheap imitations of goods designed and manufactured in more advanced countries.

Even now, according to *Scientific American* columnist Michael Shermer, 72 percent of American adults believe in angels but only 45 percent believe in evolution,[46] and Michael McGuire, professor of psychiatry and behavioral sciences at UCLA, reports that 40 percent of his students believe in haunted houses, 30 percent of American adults believe thoughts can influence the physical world and 20 percent believe it's possible to communicate with the dead.[47]

We may also hold questionable beliefs about more practical matters. Some school teachers tell children to cross streets at a corner, but if I cross in the middle of a block I need to watch for cars coming from only two directions. That makes it easy for me to see any cars that might threaten and, because I am directly in front of any cars traveling on the street, their drivers can see me. At a corner I have to watch for cars coming from four directions — one of them behind me — and because I am not directly in front of drivers who are turning, I am less likely to be seen by drivers whose path I may cross.

The city of Toronto has marked crosswalks where pedestrians have the right of way over cars. I once saw an adult 'teaching' a group of young children how to use a crosswalk on a busy street.

It seems that you can't use a crosswalk when there is no traffic. Instead, you have to wait for a car to come and then hold your hand out to make it stop, so you can cross.

The teacher was trying to teach kids how to signal drivers before they used a crosswalk, but she also taught them to not cross when there was no traffic to stop.

Some safety 'experts' recommend that people lock their car doors while driving and some cars lock their doors automatically when you put them in gear. The 'experts' tell us that locking doors will prevent them from opening in an accident but, in fact, the 'lock' on a typical

[46] Shermer, Michael, *The Believing Brain*, Times Books, Henry Holt and Co, NY, 2011.

[47] McGuire, Michael, *Believing, The Neuroscience of Fantasies, Fears and Convictions*, Prometheus Books, Amherst, NY, 2013, pg. 20.

car door just disconnects the outside handle. It can't prevent the door from opening in an accident but it can prevent bystanders from helping people in the car.

That's a sore point for me because I have personally seen the still-smoldering remains of three cars in which a total of ten people burned to death, at least partly because the doors were locked. The people inside were stunned but alive and, in the few seconds before the cars burst into flame, bystanders were unable to open the doors.

In one case, bystanders had to listen to the screams of five children as they were burned alive.

If we thought rationally about most of these things we might see the fallacies; but in our day-to-day lives we don't have time to think things out. Instead, like most other animals, we accept the beliefs of our culture.

We like to think that scientists are open to new ideas, but that belief is itself a matter of conventional wisdom. Flemish cartographer Abraham Ortelius suggested in his *Thesaurus Geographicus*, published in 1596, that the Americas were "torn away from Europe and Africa ... by earthquakes and floods," and German geophysicist Alfred Wegener developed much of the modern theory of continental drift in 1912, but the idea was laughed out of 'respectable' science.[48] As late as 1953, some physicists argued that floating masses on a rotating globe would collect at the equator and stay there, but now we know that — while he got some of the details wrong —Wegener was mostly right.

In 1858, Louis Pasteur met with violent resistance from medical doctors when he advanced his germ theory because doctors thought of him as a mere chemist, not worthy of their attention.[49] In 1859, most scientists rejected Darwin's theory of evolution and millions of Americans, including many who have been exposed to a process they call education — still reject it.

In 1879, theoretical physicist Max Planck included some new interpretations of the Second Law of Thermodynamics in his doctoral dissertation but, he complained, "I found no interest, let alone approval, even among the very physicists who were closely connected with the topic."[50]

In 1905, Albert Einstein published his *Theory of Relativity*. Soon afterward, a book was published in which 100 other scientists argued that he was wrong.[51]

In the 1920s, Prof. Ivan Wallin of the University of Colorado suggested that some species evolve by merging with other species but,

---

48 https://en.wikipedia.org/wiki/Continental_drift#Early_history

49 Vallery-Radot, R., *The Life of Pasteur*, R. L. Devonshire, trans. (Garden City Publishing Co., New York, 1926), pp. 175, 215.

50 Cited by Bernard Barber in a lecture delivered 28 December 1960 at the New York meeting of the AAAS. On the web at http://citeseerx.ist.psu.edu/viewdoc/download?doi10.1.1.392.3311&reprep1&typepdf; see also *M. Planck, Scientific Autobiography*, F. Gaynor, trans. (Philosophical Library, New York, 1949.

51 http://www.ekkehard-friebe.de/Hundert-Autoren.pdf

when the idea was ridiculed, he gave it up. Since then it has evolved into the widely-accepted Serial Endosymbiosis Theory. Biologist Lynn Margulis' first paper on the idea, "Origin of Mitosing Cells," was rejected 15 times before it was finally published by the *Journal of Theoretical Biology* in 1967.[52]

In order to develop the new varieties of wheat and produce the Green Revolution that saved millions of people from mass starvation (and won him the Nobel Peace Prize in 1970) Norman Borlaug had to battle entrenched government agronomists in Mexico, India and Pakistan.[53]

Most of us would like to believe that this kind of problem was typical of an earlier generation of scientists but, in fact, the rejection of new ideas by established scientists seems to be both routine and timeless.

In 1982, chemist Daniel Shechtman discovered a strange, five-sided crystal that did not repeat itself. He called it a 'quasicrystal' and for several years he endured the jibes of other scientists. Linus Pauling, one of the few men ever to win two Nobel prizes, said publicly that 'there is no such thing as quasi-crystals, only quasi-scientists.'

Then other scientists discovered quasicrystals in their laboratories and, in 2009, some natural quasicrystals were found in eastern Russia. In 2011, Shechtman received a Nobel prize for his discovery.[54]

For most of the 20th century, doctors around the world *knew* that stomach ulcers were caused by stress, spicy foods, and too much acid. There was no cure, but the painful condition could be relieved by a bland and tasteless diet and regular consumption of pills that could cost hundreds of dollars a month. Nobody tried antibiotics because scientists *knew* that bacteria could not possibly survive in the acid conditions of the stomach.

But in 1982, Australian doctors Barry Marshall and Robin Warren discovered that many stomach ulcers are caused by the bacterium Helicobacter pylori, but doctors around the world sneered at their discovery. Finally, to prove the point, Dr. Marshall drank a solution containing Helicobacter pylori and developed a serious case of ulcers, then cured himself with antibiotics.

Even then, for ten years the team's manuscripts and grant applications were rejected, but in 2005, the two shared a Nobel prize for their work.[55]

According to *New Scientist* magazine, some English scientists believe that the system of 'peer review' in which scientific journals allow senior scientists to review and sometimes to block publication

[52] Margulis, Lynn, *Symbiotic Planet*, Basic Books, NY, 1998, pg. 6, 53.

[53] Bickel, Lennard, *Facing Starvation, Norman Borlaug and the Fight Against Hunger*, Reader's Digest Press, 1974

[54] "Settled science is not an argument," excerpt from a speech by Dr. Arnold Aberman, May 30, 2014, printed in *National Post*, June 11, 2014, pg. FP9.

[55] Ibid.

of a paper in their field is unfair. The protesting scientists suggest that some senior scientists may slant their reviews to discredit or block rival research.[56] Even if there is no direct rivalry involved, the peer review system ensures that scientific research that supports conventional wisdom is more likely to be published than research that negates it.

Even scientists, who supposedly think for themselves, may not accept ideas until they have been approved by other scientists and printed in peer-reviewed journals. When I told one professional archaeologist (in an email) about my reservations regarding peer reviewed material, he suggested that I should write an article — and publish it in a peer-reviewed journal.

I'm not surprised that a professional archaeologist rejects my ideas because his job depends on the belief that he and others of his ilk are 'qualified' and the rest of us are not — but why would scientists ignore or discount the discoveries of other scientists? I don't pretend to explain this, but I can rationalize it.

Among most primates the young learn from and imitate the behavior of high-status adults but not low status members of their community. Author Robert Ardrey wrote of an experiment in which biologists in Japan found that only a few hours after the alphas of one troop of Japanese monkeys learned to eat wheat, the whole band would eat wheat.[57] In another band low-status monkeys were the first to eat caramels and, 18 months later, only 51 percent of the band would eat caramels. Konrad Lorenz cites experiments by Robert M. Yerkes that suggest that chimpanzees learn by imitation, but they imitate only high ranking chimpanzees.[58]

To a student at any level his or her teacher has high status, and parents and other high-status individuals expect the student to learn from the teacher. What the teacher teaches is gospel.

So when the one-time student is an adult, he will be suspicious of anyone of equal or lesser status who proposes a new idea or criticizes one that was installed by his teacher. Even if the old idea was installed by a primary-school teacher with lower status than the person who offers the new idea, it was installed by someone who was seen as high status at the time and it is being countered by someone who is seen as equal to or perhaps lower than the adult who holds the installed belief. Obviously, the belief must be considered more valid than the counter-argument.

This need not apply in every field, because many ideas are not taught in elementary schools and some fields of knowledge are completely new. I didn't learn about computers, DNA testing or

[56] "Referee or rival?" *New Scientist* Feb 6, 2010, pg. 6.

[57] Ardrey, Robert, *The Social Contract*, Dell Paperback, New York, 1970, pp. 124-129.

[58] Lorenz, Konrad, *On Aggression*, 1968 printing, University paperbacks, Methuen, London, pg. 37.

sub-atomic particles in elementary school, and I am open to new information about these topics as it comes along. Millions of people learned the story of *Genesis* in Sunday school, and they will not consider any alternative to it.

Most of us have no direct experience of most of the world and many of our beliefs are based on what we have heard, rather than what we have seen or experienced. Most of my school teachers believed, and taught me, what their teachers had taught them.

Up to a point it makes sense to trust conventional wisdom. In 1911 the English philosopher Alfred North Whitehead wrote that "It is a profoundly erroneous truism . . . that we should cultivate the habit of thinking of what we are doing. The precise opposite is the case. Civilization advances by extending the number of important operations which we can perform without thinking about them."[59]

That's true on one level because we have to make thousands of decisions a day — starting with what to wear and what to eat for breakfast — and if we had to think about all of them, we would spend most of our time wrapped up in trivia.

But it is not always a good idea to 'follow the leader.' In a jungle in Guyana, naturalist William Beebe found an army of ants marching in a circle about 1,200 feet in circumference. Each individual ant took about two and a half hours to complete the circle and thousands of ants marched for two days until most of them died.

That was in 1927, and the phenomenon has been reported several times since. It seems to happen because ants on the march follow the ant ahead of them, so if the lead ant makes a circle, it will follow the trail of the last ant in the group until most of them die. The only survivors are the few that break out of the circle.[60]

Conventional wisdom is useful but not infallible, and sometimes, I insist, we need to stop and think.

The Roots of Civilization

Even people who accept the reality of evolution may hold to pre-scientific assumptions about human origins. Many, for example, believe that when our ancestors came down from the trees they lived on a savanna. That seems to make sense because some people that we consider 'primitive' live on savannas now, but the idea is not supported by much evidence.

An alternative theory was suggested by Berlin university professor Max Westenhofer in 1942 but it was ignored by most scientists partly, perhaps, because it was seen as Nazi science. It is true that Westenhofer suggested that we are not directly descended from apes, but the operative word is 'directly.' Rather than coming out of the

[59] *An Introduction to Mathematics* (1911, ch 5, see https:// en.wikiquote.org/wiki/Alfred_North_Whitehead

[60] https://en.wikipedia.org/wiki/Ant_mill; see also Beebe, William, *Edge of the Jungle* (New York, New York: Henry Holt and Co., 1921), pp. 291-294.

trees to live in a savanna, Westenhofer suggested that our ancestors lived beside lakes or rivers or on seacoasts.

Often referred to as the 'aquatic ape hypothesis,' the idea was later outlined by English marine biologist Sir Alister Hardy in a magazine article in 1960[61] and expanded in a book by author Elaine Morgan.[62]

The trees-to-savanna scenario has no physical evidence to back it up, but the aquatic ape hypothesis was developed to explain physical evidence. Here's some of it:

We need more water than other animals. Most mammals can survive 20 percent dehydration, but humans will die at 10 percent. You might think that animals evolved on a dry savanna would be better able to survive dry conditions.

The earliest fossil considered to be human, found near the village Hadar in the Awash Valley of the Afar Triangle in Ethiopia by paleoanthropologist Donald Johanson in 1974 and nicknamed "Lucy," was of a woman who lived beside a seasonal lake, not on a savanna. Some evidence suggests that she died after she fell out of a tree but the question is where she lived, not how she died.

While most desert and savanna animals have dense hair and Arabs never go naked outdoors, some marine animals are virtually hairless with, like humans, a layer of fat under the skin for insulation. Hair is good protection from sunshine, but wet hair is not good insulation. Some marine animals — like the seal, otter, beaver and others — have hair, but they also have layers of fat.

While the nostrils of most apes would scoop up water while diving or swimming, human nostrils point down, and are protected. Human babies swim instinctively, and hold their breath under water. We are so well adapted to water, in fact, that some women give birth in water and the newborn baby swims to the surface.

Most people who have not been taught to be afraid of water can swim instinctively. I learned when I fell into a deep creek and dog-paddled to shore. Some people drown, but they are mostly those who have been taught to be afraid of water, and they panic when they fall in.

Women are the only apes with long hair on their heads. The aquatic ape hypothesis notes that this is useful for children to hang onto while their mother is swimming, and it allows them to stay on the surface if their mother ducks her head under water. When I was very young we used to swim in Lake Simcoe, near Toronto and, before I knew I could swim, I rode on my mother's back as she swam through deep water to a sandbar where I could wade.

Marine biologist Westenhofer says our spleen and kidneys also suggest aquatic development, but I don't pretend to understand this part of his argument.

---

[61] Hardy, A.C, "Was man aquatic in the past?" *New Scientist*, 7, (1960) p. 642-45.

[62] Morgan, Elaine, *The Aquatic Ape*, Stein and Day, NY, 1982.

The aquatic ape hypothesis doesn't suggest that humans actually lived in water but physical evidence shows that we are adapted to water and the hypothesis argues that we probably lived near water and were good swimmers. A 1998 BBC documentary film outlining the hypothesis included a shot of young African boys swimming in a pond and catching fish with their hands, and we know that some South Pacific pearl divers can dive to 100 feet or more beneath the surface and stay underwater for several minutes.

The most advanced ancient civilizations we know of developed along the Yellow, Indus, Nile and Euphrates rivers and 14 of the 17 largest cities in the modern world are coastal. Two-fifths of cities with populations of one million to 10 million people are located near coastlines and about half the world's population lives within 200 kilometers of a coastline.[63]

People who can't afford to live beside water often vacation by lakes, rivers or a sea, and relatively few modern people live or vacation on savannas.

While the aquatic ape hypothesis is supported by physical evidence it is generally rejected while most of the general public and even many scientists believe in the trees-to-savanna theory, which is not supported by physical evidence but is ensconced in conventional wisdom.

Conventional wisdom also holds that our early ancestors 'invented' farming before they settled in towns and villages. Archaeologists say they have physical evidence of this because geneticists tell us that wheat and other grains were domesticated about 12,000 years ago, but the oldest known city, Uruk in present day Iraq, is only about 6,000 years old.

But where were the first cities built? Most modern cities are on or near coasts and in earlier times, when ships were the only way to move heavy loads, it makes sense to assume that all cities were built on navigable water — which is usually found on or near coastlines. It also makes sense that if some of our ancestors were semi-aquatic, their first settlements would have been on a coast or by a waterway.

The ruins of Uruk — which some archaeologists think may have been the first large city — are land-locked now, but they are on an ancient channel of the Euphrates River that was navigable when the city was occupied.

Descriptions of the ruins of Uruk suggest a city so complex and well-developed that I find it hard to think of it as the 'first' city. I suggest that if archaeologists of the far future find the remains of a jet airliner — even if it is the oldest they can find — they will not declare it to be the *first* aircraft.

Archaeologists have not found the ruins of any city older than Uruk but we know that mean sea level has risen considerably — estimates

[63]http://www.prb.org/Publications/Reports/2003/RippleEffectsPopulationandCoastalRegions.aspx

range from about 400 to nearly 700 feet — in the past 20,000 years. This suggests that if the first cities were, like so many modern cities, on or near coasts, their ruins must now be under water. In fact we have physical evidence that supports this view.

In 2012, British marine archaeologists discovered evidence of a human population on "Doggerland" — the submerged land bridge that once connected Britain to Europe. On modern maps, it's shown as a shallow area of the North Sea.[64] There are confirmed sunken temples in the Gulf of Khambhat off India,[65] suspected sunken ruins off Japan,[66] and a controversial assembly that appears to be a sunken city off Cuba.[67] The Cuban site is controversial because it is too deep underwater to be explained by known sea level rise but, as imaged by sonar, it looks very much like a city.

You might think the rise in sea level would have been gradual as the ice melted and that cities would have been rebuilt above the rise. Probably some of the rise was gradual, but we know that some of it was disastrous and some may have been catastrophic.

Flash floods occurred when ice dams finally broke and released huge volumes of water in a rush. We have physical evidence of one such event in the badlands of eastern Washington State and the Columbia River Gorge in western North America which, geographers now believe, were formed by cataclysmic floods that were the result of periodic sudden ruptures of the ice dam on the Clark Fork River that created Glacial Lake Missoula. After each rupture the waters of the lake would rush down the Clark Fork and the Columbia River, flooding much of eastern Washington and the Willamette Valley in western Oregon. After the rupture the ice would reform, creating Glacial Lake Missoula again.[68]

The collapse of another ice dam released between 75 and 150 cubic kilometers of water from Lake Ojibway and Lake Agassis, which covered much of Eastern Canada, and the resulting flood scoured a huge area of land down to the bedrock that we now call the Canadian (or Laurentian) shield.

We also have evidence that continent-size glaciers have slid into the sea, causing mega-tsunamis that are hard to imagine and that would certainly have wiped out any cities on the coast and perhaps hundreds of miles inland.[69]

---

64 http://www.dailymail.co.uk/sciencetech/article-2167731/Britains-Atlantis-North-sea-huge-undersea-kingdom-swamped-tsunami-5-500-years-ago.html

65 https://en.wikipedia.org/wiki/Marine_archeology_in_the_Gulf_of_Khambhat

66 http://news.nationalgeographic.com/news/2007/09/070919-sunken-city_2.html

67 https://en.wikipedia.org/wiki/Cuban_underwater_city

68 https://en.wikipedia.org/wiki/Missoula_Floods

69 https://en.wikipedia.org/wiki/Heinrich_event. See also https://qz.com/193139/the-biggest-tsunami-recorded-was-1720-feet-tall-and-chances-are-good-it-will-happen-again/ and http://www.geosci.usyd.edu.au/users/prey/Teaching/Geos-2111GIS/Tsunami/Dominey-Howes-MarGeology07-TsunamiAustralia.pdf

Most ordinary tsunamis are caused by undersea earthquakes and are — even when they destroy coastal cities — relatively gentle. Mega-tsunamis are caused by landslides that drop huge chunks of land into the ocean, raising waves that may be hundreds or even thousands of feet high. In 1958, a mega-tsunami caused by an avalanche into Lituya Bay near the Gulf of Alaska destroyed the forest to more than 1,700 feet above sea level. If that wave had struck a relatively flat coast, how far inland would it have gone?[70]

Dozens of mega-tsunamis have occurred in the past and others — including some that could seriously damage much of our modern civilization — might occur in our time. One scenario suggests that the collapse of the Cumbre Vieja volcano on the Atlantic island of La Palma could cause a mega-tsunami that would roll right over Florida and wipe out Washington, New York, Boston and other Atlantic-coast cities. The United States might survive that, but it might not.[71]

Even mega-tsunamis pale by comparison with the theory of catastrophic pole shift proposed by engineer Hugh Auchincloss Brown in 1948[72] and revised by Prof. Charles Hapgood in his book *The Earth's Shifting Crust.*[73] Many people describe both theories as crank science but Albert Einstein, for one, took Hapgood's version of it seriously and he wrote a foreword for Hapgood's book.

In a nutshell, Brown suggested that uneven growth of polar ice caps could throw the Earth off balance and tip the planet's axis. Brown suggested that this could happen every 4,000 to 7,000 years, and that it might be due to happen soon.

Hapgood argued that there is no need for the whole planet to tip because the crust on which we live is floating on the fluid core, like the skin of an orange, and he suggests that it could slide to a new position over the core. He also argues that there was an advanced civilization before ours that was wiped out by the 'crustal shift.'

Conventional science rejects most of Brown's and Hapgood's ideas, but even conventional science recognizes some physical facts that are hard to explain without considering the possibility of crustal shift. Author Graham Hancock lists some of them in *Fingerprints of the Gods.*[74]

---

[70] Ibid.

[71]rhttps://www.cityofboston.gov/images_documents/La%20Palma%20Canary%20Island%20Generated%20Tsunami%20Study_tcm3-31980.pdf; see also http://www.livescience.com/25293-hawaii-giant-tsunami-landslides.html

[72] Plumb Robert, "Engineer Says Vast Polar Ice Cap Could Tip Earth Over at Any Time." *The New York Times*, August 30, 1948, pg. 19; see also http://www.habtheory.com/3/index.php

[73] Hapgood, Charles, *The Earth's Shifting Crust*, Pantheon Books, NY and McClelland & Stewart, Toronto, 1958. Most of it is available on line at https://archive.org/stream/eathsshiftingcru033562mbp/eathsshiftingcru033562mbp_djvu.txt

[74] Hancock, Graham, *Fingerprints of the Gods*, Three Rivers Press, NY, 1995, pp 475-486, 552-3, and 216.

One is that there are fossilized tree stumps on Mount Achernar in the Transantarctic mountains, about 500 miles from the South Pole. Even if we imagine that the whole Earth might have been warm at some time in the past, this area has no sunlight for several months every year and trees don't grow without sunlight.

Sir Ernest Shackleton found coal within 200 miles of the South Pole, and in 1935 Admiral Richard Byrd found fossils of leaves and wood on Mount Weaver, also close to the South Pole — evidence that the area was once either temperate or sub-tropical.

One of Byrd's Antarctic expeditions provided evidence that 'great rivers, carrying down fine well grained sediments' flowed in Antarctica until perhaps as late as 4000 BC. According to the report of Dr. Jack Hough of the University of Illinois: "The log of core N-5 shows glacial marine sediment from the present to 6,000 years ago. From 6,000 to 15,000 years ago the sediment is fine-grained with the exception of one layer that dates to about 12,000 years ago. This suggests an absence of ice from the area during that period, except perhaps for a stray iceberg 12,000 years ago."

A map of the world compiled in 1513 by the Ottoman admiral and cartographer Piri Reis includes Antarctica — which had not been discovered at the time — as the land is now known to lie under the ice. Reis said he compiled the map from 'ancient sources.'

Author Graham Hancock argues that Antarctica may have been the Atlantis of mythology.[75] If you find that hard to swallow, look at a globe from the bottom and you will see that Antarctica is, as Atlantis was said to be, 'an island in the middle of the world sea.' If Hapgood's theory — and Shackleton's and Byrd's observations — are valid, Antarctica may have at one time been temperate and, if so, the legendary Atlantis. I don't suggest that you should believe that, but it might be helpful to know that some people do.

There are also anomalies in the far north. Explorers have found fossilized palm leaves ten and twelve feet long and fossilized marine crustaceans of a type that could only inhabit tropical waters on the island of Spitzbergen, about half way between the northern tip of Norway and the North Pole. Again, this is an area that has no sunlight for months at a time.

Some of the frozen mammoths that have been found in Siberia had un-digested temperate-zone plants in their stomachs.

It's easy to discount the idea of an advanced antediluvian or prehistoric civilization but it's not so easy to explain why our ancestors, who were as intelligent and adaptable as we are, seem to have lived in caves and made little or no technical progress for hundreds of thousands of years while we went from clubs to computers in about 12,000. Most of us would like to think that if our civilization were destroyed we could rebuild it but we don't stop to think that few modern people could survive without modern tools and amenities.

[75] Ibid., pp 464-70

As noted earlier, the people of Tasmania did not know how to make fire when Abel Tasman reached the island in 1642. Their ancestors must have known how but somehow, after the island was cut off from Australia by the rising sea level, they lost it. If our modern technology is destroyed, we will have to start again from the beginning.

So — back to our question. Why bother with ancient pre-history? My answer is that, like sailors, we can't know where we're going unless we know where we came from; and that if our ideas about where we came from are wrong then we don't have much hope of planning a route to where we want to go.

We may not march in physical circles like the ants that Beebe found in Guyana but we do follow leaders and, because many of our so-called leaders follow leaders who followed leaders ad infinitum, we may get trapped in a web of beliefs that bears little relationship to reality.

And, like sailors at sea; if we don't know where we came from and we don't know the winds and currents that affect us, we can't possibly plan or even know where we are going.

Much of this book counters some of the lessons you probably learned in school and what you have believed most of your life. I don't expect you to accept it as gospel, but I hope it will make you think about some ideas and beliefs that you have never questioned before.

I don't demand that you accept the aquatic ape hypothesis, crustal shift, Antarctica as Atlantis or the threat of a mega-tsunami. I don't regard them as gospel myself, but I think they're worth knowing about.

And when someone asks you about the physical evidence for something in the distant past, remember that it may lie on the bottom of the sea or be, like the forests of Antarctica, buried under hundreds or thousands of feet of ice. I concede that most of my ideas are speculative but my speculations are based on facts discovered and recorded by scientists. Science may be out-ranked by conventional wisdom for some people, but it works for me.

Systems and Metasystems

Before we get into specific speculations about pre-history we need to look at some epistemological concepts that do not yet qualify as conventional wisdom. I'm going to start with an idea I developed in an earlier book[76] — the self-organized entities that I call *metasystems*.

A system can be defined as an organization that continues through a change of leadership and/or membership. The Roman Catholic Church has survived the death of dozens of popes, millions of priests and hundreds of millions of members and the United States continues

[76] Turnbull, Andy, *The System*, Red Ear, Toronto, 2005. The original is out of print but a second edition is available as a free pdf download at http://andyturnbull.com/thesystem.pdf

through regular changes of government officials and the deaths of citizens.

I got this definition of a system from the late Prof. Anatol Rapoport (one the fathers of games theory and the man who sparked my interest in systems), but for my purposes, I have to refine it a bit. In my terminology I add the conditions that a system has a formal organization and leadership, a mailing address and members who see themselves as members and who have a common purpose. We know who is pope and who is president of the United States, and we know where they live. We assume that members of the Catholic church support the church, and that American citizens support the United States.

The simplest type of human organization is a family dominated by an alpha male. In more complex societies the members of a family follow the alpha male of the family and he, in turn, follows the alpha male of the group.

This is what I call a 'charismatic leader' type of organization and it works pretty well. In ancient times Alexander the Great's armies didn't know where they were going but they followed Alexander to conquer countries they had never heard of before.

If a charismatic leader dies but the members choose to follow the course he set, rather than a new leader, the organization may morph into a system. Alexander built an empire but it did not morph into a system, and it broke up soon after he died.

The followers of a charismatic leader serve the leader but members of a system serve a chosen or assigned purpose and the leader can lead only in the direction of that purpose. The Pope can't make Catholics worship the devil, the President of France can't crown himself king and the president of the East Idaho Ladies Sewing Circle can't make the ladies turn from sewing to acrobatics. These organizations have their path set and the function of the leader is to lead them on that path.

If a system works well, the members will value it and they may consider the welfare of the system to be important. As time passes and circumstances change the original purpose of the group may become secondary and the survival of the system more important. Eventually the original objective may be lost or altered, but if the system itself serves a function it may survive.

In 1981 a group of truckers in the village of Notre Dame du Nord, Quebec, organized an uphill drag race of loaded trucks to raise money to help build a new arena. The arena was completed long ago, but the Rodeo du Camion is now an annual event that attracts entries from across North American and raises millions of dollars for civic projects in and around the village.

The group that ran the original event has evolved into a system that continues to stage the races because people like them, they raise money for civic projects in and around Notre Dame du Nord and they provide full or part-time jobs for dozens of villagers. When the original purpose was accomplished, the system found a new purpose.

We might question whether a commercial corporation can outgrow its function because that function is, ultimately, to make money; but there is no question that it can evolve. The Minnesota Mining and Manufacturing Co., generally known as 3M, was formed in 1902 to mine abrasives to make grinding wheels. The first mine was a failure but 3M went on to sell about $30 billion a year worth of products we see or use every day. Two of its best-known products are Scotch Tape and Post-it Notes, neither of which have much to do with mines, abrasives or grinding wheels.

At this point both mining and manufacturing are incidental to the entity we call 3M. The business of the corporation is to survive and grow, and any single commercial activity is secondary.

If and when the survival of a system organized and nominally controlled by humans becomes the primary objective of the members, the system may become a near-autonomous entity. In this form it looks very much like other systems but where some systems serve the humans who manage them, others manage the humans who serve them.

During World War II, economist John Kenneth Galbraith served as deputy head of the U.S. Office of Price Administration, with authority to set prices for many commodities traded in the United States. The final decision was his but he was aware that no single man knew enough to make such decisions and that the best he could do was to accept the advice of the bureaucrats who served under him. In his words,

> Decisions on prices — to fix, raise, rearrange or, very rarely, to lower them — came to my office after an extensive exercise in group decision-making in which lawyers, economists, accountants, men knowledgeable of the product and the industry and specialists in public righteousness had all participated. Alone, one was nearly helpless to alter such decisions; hours or days of investigation would be required and, in the meantime, a dozen other decisions would have been made. Given what is commonly called an 'adequate staff' one could have exercised control. But an adequate staff would be one that largely duplicated the decision-making groups with adverse effects on the good nature and sense of responsibility of the latter and even more time required for decision. To have responsibility for all of the prices in the United States was awesome; to discover how slight was one's power in the face of group decision-making was sobering.[77]

Galbraith was the head of his office, but the office itself made the decisions. If he had the 'adequate staff' that he mentions, it would

[77] Galbraith, John Kenneth, third edition of *The New Industrial State*, New American Library, NY, 1979, pg. 60.

still have been the staff, rather than Galbraith himself, that made the decisions. With or without an adequate staff, the titular head of a system is a servant of the system, not its master.

He may not even be a key player. Galbraith notes that the director of the Manhattan Project that developed the atomic bomb during World War II was U.S. Army General Leslie R. Groves. As the general in charge of a huge project Groves travelled by limousine. Physicist Enrico Fermi, who played a key role in the development of the bomb, rode to work on a bicycle.

If Groves had quit, he could have been replaced the next day by any of thousands of other men. Whoever he was, the replacement would also have travelled by limousine and demanded the impressive office and other perks that General Groves enjoyed.

If Fermi had left, the project would have ground to a halt. Groves might have kept his job and his limousine and his office but, without Fermi, he could have accomplished nothing.[78]

The man who runs a large organization may be an exceptional person but, management consultant Peter Drucker says, almost anyone could serve as the head of a well-organized system. In *Concept of the Corporation* he wrote:

> No institution can possibly survive if it needs geniuses or supermen to manage it. It must be organized in such a way as to be able to get along under a leadership composed of average human beings.... No institution has solved the problem of leadership, no matter how good its formal constitution, until it gives the leader a sense of duty, of the importance of his trust and a sense of the mutual loyalty between him and his associates; for these enable the average human being — and occasionally someone well below average — to function effectively in a position of trust and leadership.[79]

For a sobering view of the nature of leaders and leadership, consider the results of an experiment by physiologist Erich von Holst of the German government's Max Planck Institute, as described by zoologist and philosopher Konrad Lorenz.

Fish that swim in schools prefer to stay near the center of the school, where they are safe from predators. Schools of fish don't often travel fast or far, because fish on the edge of the school keep trying to move toward the center.

Some fish leave the school to chase after food but they return immediately after they catch it. If there is a lot of food in one direction the school will move that way because as some fish return to the school others leave it to obtain the food, but there is no concerted movement.

---

[78] Ibid., pg. 77.

[79] Drucker, Peter, *Concept of the Corporation*, John Day Co, NYC, revised ed. 1972, pp. 26-27.

In his experiment von Holst removed the forebrain — which controls the schooling instinct that pulls fish to the center of a school — from a common minnow.

That minnow became the undisputed leader of the school. It would swim toward food or anything else that interested it, and some fish would always follow. Because the 'leader' lacked a forebrain it never felt the need to turn back to the center of the school and, because it never turned back, the others followed it.

The function of a leader is simply to lead and, in many situations, any direction, right or wrong, is better than no direction. For some systems the ideal leader may be the kind of person that Russians used to call an apparatchik. This is a bureaucrat or other insider who may have no technical competence or leadership ability at all, but who has friends and knows the rules and works both to his advantage.

A typical apparatchik will take great care to see that all rules and procedures are followed to the letter. He may be a dead loss as a manager but still a good leader because, with rules to follow, he will have no doubts. A brilliant scientist might not be a good leader because he would question even his own decisions.

In either case, the system is more important than the leader.

Metasystems

A metasystem may develop when two or more entities have a common purpose or problem, or when the behavior of one affects or enables the behavior of another. They behave like systems but they are self-organized and they may include competitors or predators who prey on other members of the same metasystem. Predators and prey are not allies but, because they affect one another's behavior, they form a metasystem.

Metasystems are natural developments found in relationships between life forms around the world. Some are formed by co-evolution, when organisms have evolved together to depend on one another.

An insect called the fig wasp lays its eggs only in one type of fig, and that type of fig is pollinated only by fig wasps. If there were no wasps, the figs could not reproduce and, without figs, the wasps could not reproduce. Yucca plants are pollinated only by yucca moths, which hatch from eggs laid only in yucca plants. Co-evolved life forms depend on each other for survival.

A different type of metasystem, which biologists describe as emergent behavior, develops because it benefits its participants. In Arabia a thorny-tailed lizard may share a burrow with a black scorpion. The lizard eats insects and the scorpion could kill the lizard, but when they live together, the lizard eats predators that might attack the scorpion and the scorpion protects the burrow — and therefore the lizard — from foxes and men that hunt lizards. Either can live without the other, but both benefit from having a partner.

In African rivers Egyptian plover birds hop into crocodiles' mouths and pick food scraps from between their teeth,[80] and on reefs around the world wrasse fish clean the teeth of sharks and barracudas. Other types of fish clean the teeth of hippopotami[81] while still others groom the huge beasts. Oxpecker birds pick ticks off African cattle, and warthogs in Uganda lie down and roll over so banded mongooses can groom them and eat the ticks that bother them.[82]

In Brazil, dolphins herd schools of fish up to human fishermen, signal the fishermen when to throw their nets, then catch the fish that escape the nets. Biologists have observed a mother dolphin teaching her calf how to do it.[83]

Fishermen and dolphins cooperate because both catch more fish when they work together. Similar collaborations between dolphins and human fishermen have been observed in Australia and Africa.

These relationships are all self-organized and, because they are self-organized and have no management, autonomous. I see them as entities in their own right.

We might compare the members of a metasystem to an ant colony, which many biologists consider to be a single organism. As far back as 1911 biologist Wm Morton Wheeler wrote (in the *Journal of Morphology*) that "the ant colony is an organism." In modern times biologist Edward O. Wilson describes an ant colony as a "diffuse organism."[84]

The colony consists of individual physical organisms — ants — and a metaphysical[85] something-or-other that ties them together. In the same way a metasystem consists of the humans that serve it and the metaphysical component that unites them. It's easy to assume that the organizing force in an ant colony is 'instinct,' but instinct is internal and while it can guide an individual ant, it can't rule a colony. We might question whether the metasystem is inside or outside individual ants, but there is no question that the ants are inside the metasystem. I concede that the metasystem could not exist without the ants and their instinct, but neither could the colony.

Metasystems have been part of humanity for a long time and I suggest that what we think of as human history is actually the history

[80] https://www.youtube.com/watch?vDd6GcQrkMDM

[81] http://feedthedatamonster.com/home/2014/5/14/mutualism-of-the-month-hippopotamus-and-their-fish-partners

[82] https://www.sciencedaily.com/releases/2016/03/160307150359.htm

[83] Simoes-Lopes, Paulo C. with Marta E. Fabian and Joao O. Menegheti, "Dolphin interactions with the mullet artisanal fishing on southern Brazil, a qualitative and quantitative approach," *Revista Brasileira Zoologica* 15(3), 1998, pp 709-726.

[84] Wilson, Edward O, *The Insect Societies*, Belknap Press of Harvard University Press, Cambridge, MA, 1971.

[85] The word 'metaphysical' need not refer to ghosts or spirits. It describes entities that exist, but have no physical form. The English language is metaphysical, and so are customs, traditions, 'the market' and dozens of other things that we are familiar with.

of metasystems and, to a lesser extent, systems. I argue that they produced the civilization that we see as a human accomplishment, and that they control it. Some metasystems have robbed most of us of our basic human nature and now they threaten the very future of humanity as we know it. Even though they have no physical form some metasystems have considerable physical power and, as we will see, they do not all serve humanity.

Probably the best known metasystem is the Military Industrial Complex, AKA the MIC, which is the informal and often unacknowledged alliance between military forces and the industries that supply them. Some politicians; some members of the MIC and even some members of the public pretend that there is no alliance but it is obvious that military forces could not function without arms-makers to produce their weapons, and that arms-makers would not be profitable if there were no military to buy their products. If they depend on one another, they are allies.

Most of us see the MICs of different nations as discreet entities but, as American Secretary of State, Henry Kissinger recognized an unholy alliance between the MICs of NATO and the Warsaw Pact.

In *Almost Everyone's Guide to Economics*, authors John Kenneth Galbraith and Nicole Salinger wrote that American Secretary of State Henry Kissinger "once told me that you could understand the relations between the Soviet Union and the United States only if you realized that the proponents of military expenditures in both countries had united against the civilians of both countries."[86]

Kissinger realized that the supposedly-opposed MICs were actually allies, because development and growth of the Russian military establishment drove growth of the American MIC and vice versa. The cold war was a burden for Russian and American civilians but a boon to the MICs of both countries.

In the modern world the so-called 'War on Terror' is in fact a War of Terror in which Western military, police and security forces are allied with Islamic terrorists against citizens of Europe and the United States and Muslims everywhere; and the 'War on Drugs' is an alliance of American police forces and the Drug Enforcement Administration with Mexican cartels and American drug dealers, each side supporting the growth and prosperity of the other.

If there were no drug dealers, the police and the DEA would not be as well funded as they are. If police and the DEA did not try to stop the drug trade, drugs would be cheaper and the trade not so profitable.

I do not suggest that the government agents who profit from these wars are deliberately malign, but I do argue that, in many cases, they do more harm than good.

Because they are self-organized, metasystems are autonomous and they don't need any human leaders or even recognition. In 1986,

86 Galbraith, John Kenneth and Nicole Salinger, *Almost Everyone's Guide to Economics*, Houghton Mifflin, Boston, 1978, pg. 145.

computer programmer Craig Reynolds wrote a program to control the behavior of virtual birds, which he called boids. He found that if his boids were given only three rules they would behave very much like a real flock of birds, or a school of fish.

Aside from the obvious — that it must avoid colliding with fixed objects in its virtual environment — each boid controlled by Reynolds' program is required to travel in the same general direction as other boids and to try to stay near the center of the flock but avoid crowding other boids.

Each boid 'knows' where all the other boids are but reacts only to the boids in its immediate neighborhood.

Reynolds' three rules can control a flock of thousands of boids or other virtual animals and the behavior they produce is very similar to the behavior of real animals. A modified version of this program controlled the behavior of swarms of animated bats and flocks of animated penguins in the 1992 movie *Batman Returns* and has been used in other films, TV commercials and demonstrations.[87]

Birds don't know Reynolds' rules but some will have a natural propensity to behave one way and some to behave another. Birds whose natural propensity is to behave as Reynolds' rules demand form flocks, and birds that do not, do not. Flocking develops and survives because birds that flock are more likely to survive than birds that do not. Birds near the center of a flock are generally safe from raptors because a raptor can't dive into a flock without risking a collision that might injure it. Birds that leave the flock and fly on their own are easy prey, and so are birds that fly at the edges of the flock. Birds that do not avoid fixed obstacles or other birds will collide and be injured or killed.

Because it improves the chances of survival, flocking behavior is sustained from generation to generation and, even though metasystems have no recognized leaders and no formal structure, they may be more stable than any formal system.

Birds are probably not aware of the rules that control their flocking and humans — even people who serve metasystems — may not be aware of the metasystems that control them. The one thing that all members of a metasystem have in common is that the welfare and/or the behavior of any member can be affected — whether consciously or not, intentionally or not — by other members of the metasystem.

As a general rule, the smaller the metasystem, the closer the bond. Human beings are all life-forms and members of the animal kingdom and of the same species but, beyond that, we are also members of one race or another, we are all have a religion or lack one, we all speak one or more languages, are citizens of one or more countries and are members of social and occupational groups. My bond with white skinned English-speaking writers is stronger than my bond with, for example, a North Korean soldier or an African elephant-poacher, my

[87] www.red3d.com/cwr/ boids/

bond with any human is stronger than my bond with any non-human life form and my bond with any mammal is stronger than my bond with an insect.

Neither systems nor metasystems have any bond with humans or concern for our welfare. We know that smoking is a serious health problem but, long after the danger of smoking was recognized, tobacco companies added nicotine to cigarettes to make them more addictive. Men and women who had children of their own planned and used advertising to attract teen-agers who were not yet addicted and, in effect, they encouraged their own children to smoke.

We know that our use of automobiles strangles cities, creates health problems and contributes to climate change that may cause a global famine but, under the influence of the metasystems of the automobile and oil industries, the governments we elect to protect our interests enforce laws that demand 'emissions controls' — which reduce the efficiency of gas engines and increase both fuel consumption and total emissions — rather than the simple and obvious option of limiting or taxing engine size.

Most of the most fuel-efficient cars in the world are not sold in North America because they do not meet American 'standards.' Nearly 30 percent of the cars in Japan are the class they call 'mini-cars,' with engine displacement of 650 cc or less and which burn about half as much fuel as the smallest cars sold in the United States; and some countries approve the use of cars with tiny lawnmower-size engines. The smallest cars are not allowed on expressways, but they are very efficient in cities and can be used on rural roads.

Most North Americans think they need big engines to travel the long distances of our large countries, but that's an illusion. For a couple of years in the 1960s I was a free-lance writer, photographer and TV camera-man based in Owen Sound, Ontario, covering a radius of about 100 miles around that city for three newspapers, two TV stations and an assortment of magazines in my Citroen 2CV car, which had a 425cc engine — little more than a quarter the size of the engine in a typical American 'small' car. For about three years in the 1970s I lived in Belleville, Ontario (about 110 miles east of Toronto), and wrote an average of one article a month for an Alberta government magazine based in Edmonton, more than 2,000 miles from my home. I drove my Toyota Corolla, with a 1,200 cc engine, from Belleville to Alberta at least once a month. I later drove that car to Inuvik, in the Northwest Territories — about 4,500 miles from Belleville — and about 50 miles out on the ice of the Beaufort Sea off Canada's North coast.

Many small cars around the world have diesel engines, which use very little fuel and therefore produce very little exhaust while idling. This makes them ideal for use in cities — where cars spend a lot of time idling — but they do not require as much maintenance or replacement of spark plugs and other parts as gas engines. American emissions regulations discriminate against diesel cars because, even though most of them have overall emissions lower than any gas-engine car sold in the United States, they do emit some nitrogen dioxide, which is a no-no in the States.

Some governments pass laws that will harm their citizens. In the summer of 1998 the government of Canada changed a law to permit the use of manganese methylcyclopentadienyl tricarbonyl as an additive in gasoline. MMT, as it's usually called, is illegal in some US states because it is known to cause damage to human nervous systems. It was illegal in Canada until the American manufacturer threatened to sue the Canadian government, and Canadian law was changed to allow it.[88] The families of the politicians that approved the change, and of the oil company executives that sued for it, are among the people whose nervous systems are being damaged.

We know that gambling is addictive and that it does serious social and economic damage but state and provincial governments across North America license or run casinos and manage or allow lotteries. Even if the politicians and bureaucrats are not harmed themselves, the gambling they encourage harms some people and may destroy their own families.[89]

Many metasystems include predators who prey on other members of the same metasystem. Predators and prey are not allies but, because they affect each other's behavior, they are part of the same metasystem.

A chief of police is not likely to join the American Bar Association and the National Association of Chiefs of Police would not accept a drug lord or a gang of bank robbers as members but, because their occupations intersect, many lawyers and all policemen are part of a metasystem that includes drug lords, bank robbers and other criminals.

Systems are mortal because they are organized and managed by people, and, in many cases, the people who organize and/or manage them live comfortable lives. Because they are comfortable, they don't want change and because their leaders don't allow change, systems don't adapt — or don't adapt well enough — to changing conditions. humans have established countless kingdoms and empires but none has lasted more than a few hundred years. Because a metasystem like the MIC is self-organized, it has no management to prevent change and, while parts of it may die, the metasystem itself will adapt to changing conditions.

---

88 McArthur, Keith, "NAFTA ruling goes against Ottawa," *The Globe and Mail*, Nov 14/00, pg. B1.

89 For an outline of the kind of harm government-sponsored gambling does to our society see the article "We're all paying for Government's Gambling Addiction," by Bruce Hutchinson, starting on page 47 of the April/97 edition of the Canadian edition of *Reader's Digest*. For a specific example of a municipal official who helped bring gambling to his town, was hooked himself and wound up embezzling from trust accounts to support his addiction read *Losing Mariposa* by Doug Little, ECW Press, Toronto, 2002.

The first MIC must have formed thousands of years ago and MICs have been forming and merging ever since. The systems that spawned the first MICs are long-gone but — because there has never been a time without one — we must consider the global MIC that exists today to be a development, rather than a descendant, of the original. As a system can survive the loss of individual members and even leaders, a metasystem can survive the loss of individual members and a compound metasystem can survive the loss of individual metasystems. Even in ancient times, the loss of a Roman Legion or the conquest of an enemy meant little to the compound MIC of the day, because Rome had more legions and lots of enemies.

In our time the global MIC has grown to enormous size because nations have formed larger and larger alliances — like NATO, SEATO and the Warsaw Pact — and each one merged the MICs of the countries within it. With a bigger industrial sector, each MIC is able to develop new weapons, driving rival MICs to develop new defenses and more new weapons.

The global MIC is enormous and virtually immortal but metasystems don't have to be big or long-lived. When I ride an elevator, I share with the other passengers an interest in the quality or lack of music, in the freshness of the air and in the hope that the mechanism will not break down — but only for a few minutes. When I leave the elevator, the metasystem continues — because the occupants are still concerned about the state of the elevator — but I am no longer a member of it. When the elevator is empty, the metasystem dissolves.

Real Life Examples

When I first wrote it I thought the example of the metasystem in the elevator was hypothetical, but it turns out to be reality. In *The Social Animal*, psychologist Elliot Aronson cites the famous case of Eleanor Bradley, who tripped and broke her leg while shopping on Fifth Avenue. She lay on the sidewalk, begging for help, for 40 minutes while hundreds of people walked by; but none helped her or called an ambulance.

The New Yorkers who bypassed Eleanor Bradley had no common bond but, where a bond exists, people will help. Aronson also wrote of the time he was camping in Yosemite National Park and he heard a man's voice cry out in the night. He grabbed a flashlight and went looking, and saw dozens of other lights heading the same way. It turned out that the problem was not serious — a man shouted, partly in surprise, when his camp stove flared up — but the example is interesting.[90]

The people who bypassed Eleanor Bradley would not have risked any danger if they offered help, but if the man's cry in Yosemite had

[90] Aronson, Elliot, *The Social Animal*, Viking Press, NY, 1972, pp. 38-40.

been due to a bear attack, the campers who ran to help might have been in danger.

Aronson speculates that because the people in the park were all camping, they felt some kind of bond.

The effect, if not the reason, was demonstrated by an experiment by Irving Piliavin[91] in which an accomplice staggered and collapsed in a New York subway train. He lay still and stared at the ceiling and, in most of 103 trials people rushed to his aid. Even when the actor carried a bottle and reeked of alcohol, he got immediate help in about half the trials.

Aronson suggests that people in the same subway car "have the feeling of sharing a common fate." I suggest that they form a metasystem.

Our membership in some metasystems — as a passenger in a train or an elevator or a camper in a park for example — is brief, but our membership in others — such as gender and racial groups — may be for life.

The Nature of Systems and Metasystems

Once formed a system or a metasystem may evolve and grow like a life-form, but is it really alive? We might argue that but, according to author Ellen Thro, European computer programmers working on artificial life have agreed that when a program becomes autonomous it is alive.[92] Because it is self-organized and has no human managers a metasystem is autonomous and therefore, by the standards of artificial life, alive.

And whether they are really alive or not, if we think of both systems and metasystems as life forms we remind ourselves that they have their own wants, needs, intentions and ambitions — all of which are distinct from those of the humans who may think they control them.

Of course we have no reason to believe that either systems or metasystems actually want, need or intend anything, but here I use the words loosely. We know that flowers can't possibly want or like sunlight, but many flowers grow better when they have sunlight and some will turn to face the sun. In the vernacular we say that these flowers want or like sunlight, even though we know they cannot possibly want or like anything. When I say that a system or metasystem wants or likes this or that, I do not suggest that it is conscious.

Neither do I suggest that either systems or metasystems can have intentions as we normally use the term, but here again I use traditional license. Biologists know that the bright colors and nectars of flowers

---

91 Ibid., pg. 43; see also https://prezi.com/c25l7fq_zxpb/piliavin-et-al-good-samaritan-study/ and http://www.tutor2u.net/psychology/reference/piliavin-1969

92 Thro, Ellen, *The Artificial Life Explorer's Kit*, Sams Publishing, Carmel, IN, 1993, pp. 5-15.

are not *designed* to attract bees. They just happened, and because they attract bees, the flowers that evolved them have flourished. By convention, we attribute *design* and *intent* to many natural phenomena and in the same way, I attribute intent to the different types of systems even though they can have none.

It makes no difference to a fly whether a Venus flytrap flower intends to trap it or not. Either way, the fly is trapped.

Neither the flower nor a metasystem need intentions. They do what they do because they are what they are. Their 'behavior' may have developed by random mutation, but it survives because it helps the entity to survive and grow.

Even animals don't *want* to grow. Evolution does not care what plants, animals or things want. Changes occur, for one reason or another, and the ones that work are passed on.

Within human society a system or metasystem can grow by gaining control of more money, power or people. Most of the humans involved in them would like to see the commercial systems we call corporations make profits but the systems themselves want only to grow. We see this when huge corporations continue to grow even though they continually lose money.

Once established, most corporate systems tend to grow because the humans who manage them are ambitious and if they can't find a legitimate need for a new product, service or level of administration, they can rationalize one. As the corporation's needs change some products, services and administrations may become redundant but they will not be discontinued if the systems or metasystems that supply them find some way to maintain the appearance of need, if not the need itself.

If that sounds unreasonable remember that I speak of effects, not causes. Some metasystems will grow and some will fail but it is only the ones that grow that we need concern ourselves with. They are the evolutionary successes and the others are failures.

A successful metasystem will even defend itself. Faced with the same stimulus some metasystems will respond this way and some that, and some will survive and some will not. The survivors are those whose response to a threat amounts to an effective defense.

At times one system or metasystem may cooperate with another but, whether cooperative or not, human systems tend to be competitive. This may be partly because they are potentially cannibalistic. The fastest way for one system to grow is by absorbing other systems and we have seen this occur throughout history.

In years gone by empires grew by conquering nations and other empires and, in the past hundred years, we have seen many of the commercial systems we call corporations grow to enormous size by taking over other corporations.

In the modern world nations do not take over other nations completely but they do conquer them and set up puppet governments, integrating the conquered nation into the conquering system if not the conquering nation. In the Second World War Nazi Germany set up the Petain government in France and the Quisling government in Norway, and the governments of Iraq and Afghanistan are now supported by foreign troops.

In most cases the citizens of puppet states are not happy but, to a system or metasystem, human considerations are of no more interest than the long-term interests of cattle or other animals are to a farmer. Systems and metasystems are not opposed to humans any more than humans are opposed to cattle but, when the interests of metasystems and of people clash, metasystems will support the interests of metasystems and the servants of a system will support the system they serve.

A new leader might change the nature and behavior of a system, but that is not likely. Over time a system, like any living thing, develops a nature in response to internal and external pressures. Within limits the leader can steer a system as the captain steers a sailing ship but, like the captain of a sailing ship, the leader of a system can steer only in directions that conditions — including the nature of the system — allow.

Like the captain of a sailing ship, the leader of a system can beat against the wind; but that takes time and, unlike a sailing ship, a system may be autonomous. In many cases a system will choose the individual who leads it and, most times, it will choose a leader who is committed to the course the system is already following. Some systems will resist or get rid of a leader who tries to change it. As we noted earlier, the Pope could not make Catholics worship the devil.

We might question whether a system or a metasystem uses people or whether the people it uses are part of the system or the metasystem but I'm going to bow out of that one. If physicists can say that light is a wave motion that sometimes shows the characteristics of particles, or a shower of particles that seem to travel in waves, I can fudge on the exact nature and composition of a system or metasystem. My personal feeling is that some people are part of systems or metasystems and others are used by them.

Compound Metasystems

One metasystem can include another, and another and another in a seemingly endless chain. A termites' nest or termitary, for example, may be 20 or 30 feet high and 10 or 15 feet in diameter and it often goes as far down into the ground as it rises above it. It has an underground water supply and an elaborate ventilation system. According to entomologist E. O. Wilson, the nests of Macrotermes natalensis maintain the temperature of their internal fungus gardens within one degree of 30C and the carbon dioxide concentrations about 2.6 percent, with very slight variation.[93]

The outer shell of a termitary is made of grains of sand cemented together by a material the termites secrete, and the interior is a cardboard-like structure made of their excretions. The structure is waterproof while the colony lives — which could be more than 50 years — but it will start to crumble less than a year after it is abandoned. South African naturalist Eugene Marais compared the individual termites in a colony to blood cells, and the termitary itself to the exoskeleton of an insect.[94]

We see each individual termite as a single entity, but that's an illusion. Termites eat wood, straw and other cellulose-rich food but they can't digest cellulose. The enzymes that do it for many Australian termites are produced by a protozoan called Mixotricha paradoxa, which swims around in the termites' gut. It's an active little critter and it moves around a lot, but not under its own power.

Many protozoans swim by waving hairlike projections called flagella, which act like oars, but Mixotricha paradoxa doesn't have enough flagella to do the job. Fortunately, lots of the type of bacteria that biologists call spirochetes attach themselves to the surface of Mixotricha paradoxa and wave like flagella to move the protozoan around.

The other surprise is that Mixotricha paradoxa probably can't digest cellulose, either. Fortunately, it plays host to at least two symbiotic bacteria that do the job for it.[95] The bacteria enable the Mixotricha paradoxa to produce the enzyme that digests the cellulose that the termite eats and thus enable the termite to build the termitary.

Each individual termite is a system with internal metasystems that have internal metasystems and, at the same time, is itself part of a larger metasystem. We could also say that individual human beings are metasystems because we also carry bacteria that digest the food we eat, but the human systems and metasystems we will consider in this book are all composed of complete humans. Most of them are groups of humans, but some human-based systems and metasystems include animals or machines.

A cavalry regiment could not exist without horses and an air force could not exist without airplanes. Some military units exist without either horses or airplanes but an infantry regiment is very different from either an air force command or a cavalry regiment.

---

[93] Wilson, Edward O, *Sociobiology*, Harvard University Press, Cambridge, Mass, 1975, pg. 11-12.

[94] Eugene Marais' work *Die Siel van die Mier* was originally published in Afrikaans, in 1925. The English translation, *The Soul of the White Ant* was published in 1937, after Marais' death, by Dodd, Mead & Co, NY.

[95] Margulis, Lynn, "Symbiosis and evolution," *Scientific American*, 225(2) pp. 49-57, 1971.

The horses make the difference between cavalry and infantry and that difference may be very real. A troop of cavalry might have more in common with, and possibly more respect for, a group of civilian fox hunters or a team of polo players than with and for a company of infantry. Some troops of cavalry may include men who are also members of a fox hunt or a polo team, but neither a cavalry troop nor a fox hunt would include infantrymen.

Some human systems also include places. A city council, for example, could not exist without its city.

Wheels Within Wheels

As we saw among termites, discreet entities may form metasystems that are in turn components of other metasystems or systems. We like to think that corporations, governments and other systems are under human control but the fact is that most are run, directly or indirectly, by internal metasystems. The 'inner circle' may be a clique of officials — part of the sub-system that Galbraith called a "technostructure" — and this explains why a change of government may not bring a change in policies.

As Galbraith observed, most leaders' decisions have to be based on the research and recommendations of a technostructure and, in most cases, the technostructure does not change when a new government is elected. Most of the technostructure that delivered information to Barack Obama, for example, was essentially the same organization that informed George W. Bush; so it should be no surprise that two presidents with different attitudes made similar decisions.

Governments may also be influenced by people who have no official connection with them. Many wealthy people and corporations, for example, offer campaign contributions to two or more political parties and/or candidates. Whatever their own beliefs, they consider it good business to buy access to the inner circles of all political parties so they are covered no matter which wins election.

This can extend beyond national boundaries. We have already noted, for example, that national MICs around the world are united in a global MIC. We assume that each individual member of every national MIC is loyal to the system he serves but, because each MIC is a metasystem with no fixed membership and because all MICs have common interests, members of each national MIC must also be members of the global MIC which, because it thrives on tension and threat of war, must be seen as an enemy of every individual nation.

Ultimately, we are ruled by metasystems and, whether human or not, metasystems are natural phenomena. The organization of an ant hill is produced by the behavior of the ants, but that does not make it less natural.

Systems and metasystems both control their members, but the level of control varies. Among the insects — such as ants and bees

— that biologists describe as "eusocial," the metasystems' control is absolute. Some human systems, such as Notre Dame du Nord's Rodeo du Camion, are run mostly by volunteers and the demands they can make on their members are limited. Working for a commercial company like 3M is a career for tens of thousands of people and the company can move many of them around the country to suit its needs. The systems we call armies and churches, and terrorist groups like Al-Qaeda can send men to their death.

Control does not have to imply that either systems or metasystems have intelligence or intentions as we understand them. Obviously a metasystem could have neither — but our understanding of these things may be open to question. Some economists believe that the metaphysical entity they call "the market" is self-regulating, and some things that have no brain can behave in ways that we would normally expect of an animal with intelligence.

Plasmodial slime molds develop from spores that settle on soil, leaf litter or rotting wood and morph into individual amoebae.

For a while, the amoebae creep through films of water toward any source of light, and they engulf bacteria like other amoebae as they move but, in time, the random population of amoebae coalesces into a blob called a plasmodium that can resemble a small worm or a disc a couple of inches across. It holds this form for about two weeks — still creeping through films of water and toward light — then transforms itself into a plant shape with a base and stem supporting a top that will produce spores that will develop into new slime molds.

Slime molds have no brains or discernible nervous systems, but they can find their way through a maze. At the Bio-Mimetic Control Research Centre in Nagoya, Japan, scientist Toshiyuki Nakagaki laid out a 3x3 cm maze pattern, with two outlets and four possible routes from one to the other, on a sheet of agar.

He placed small piles of ground oat flakes at both outlets of the maze and pieces of Physarum polycephalum slime mold in it. Initially, the slime mold spread out to fill the whole maze, but, after eight hours, it contracted into a tube that filled just one route — the shortest one — between the two piles of food.[96]

We have no reason to suspect that slime molds might be intelligent, but at least they are alive. What about a drop of oil, dyed pink and doped with acid?

Northwestern University chemist Bartosz Grzybowski put one such drop on a film of water at the entrance to a complex maze and found that it followed a reasonably direct route to reach an acid-soaked lump of gel at the exit. In fact, it did better than the slime mold because the mold filled several dead-end passages and then had to withdraw from them. The doped oil found the right route with no mistakes.

[96] Nakagaki, T. et al. "Maze-solving by an amoeboid organism," *Nature*, 407: 470 (2000).

Sounds weird? Yes, but there is an explanation. The report I read explains that acid from the gel seeps through the film of water, making it more acidic the closer you get to the gel. The acid in the oil changes the surface tension of the droplet, attracting it to more acidic sections of the maze and eventually to the gel at the exit.[97]

It's not intelligence as we understand it, but it looks very much like the kind of behavior that can be driven by intelligence. I suggest that both systems and metasystems can also display behavior and I think of them, and of the global metasystem that I call *The System*, as living entities. It may be a stretch to think they are alive but there is no doubt that both exist and that they can behave as though they were alive. Whether they are actually alive or not is a matter of semantics and I leave that for another argument.

Living on Autopilot

It's all right for insects and slime molds to behave like robots but human beings are intelligent and we make our own decisions, don't we? Well, sort of.

We are rational and we can make our own decisions, but we don't think about everything we do. We can't, because life and the world are too complex for conscious thought.

As I write this my reason is concentrated on the argument and the operation of my computer. At the same time my autonomic nervous system handles the much more complex tasks of maintaining my heartbeat, breathing, body temperature and other functions. It does these jobs with no help from my conscious mind and, in fact, my conscious mind could not control many of them. The signals that control them come from my brain, but my conscious mind is not involved.

When I walk down the street learned responses control my balance and the mechanics of walking, and maintain automatic collision-avoidance systems to keep me from bumping into lamp-posts and other pedestrians, even though I may not consciously see them as obstacles. Walking is a learned skill and I know how to avoid obstacles but I don't have to maintain conscious control very often. My brain does that without much help from me.

Other automatic systems are alert for the sound of brakes, squealing tires, running feet and other audible or visual cues that could signal a threat to my safety. I am vaguely aware of motor traffic, road crossings and other potential hazards, but I don't really think about them. If an emergency does occur, my automatic systems will usually respond to it before my conscious mind knows what is happening.

---

97 Barras, Colin, "What a maze-solving oil drop tells us of intelligence," *New Scientist* #2744, 21 January 2010. On the internet at http://www.newscientist.com/article/ mg20527443.900-what-a-mazesolving-oil-drop-tells-us-of-intelligence.html

In unfamiliar territory I may think about my route but on my home ground I may choose a destination when I set out and not think about it until I arrive. Sometimes my automatic navigation system may take me to the wrong destination, as when I set out for a store that I seldom visit and walk past it to a familiar library.

If we actually thought about everything we did it would take us all day just to choose a salad for lunch. Instead we form preferences and we usually ignore other possibilities.

Our preferences are based partly on experience and partly on information we accept from others. I know from experience that I like a Caesar salad but if you recommend a Waldorf and I trust you, I might try it.

Most of us rely on rehearsed or prescribed behavior in most situations. In our social intercourse we know what 'is' and what 'is not' done; in our personal lives we have 'habits' and, if we work for a system, there are appropriate 'procedures' to follow for most situations.

I don't know enough mathematics to calculate the path of a thrown ball but, even if I see only a small part of its arc, I can usually catch it.

We may prepare for physical activity with mental rehearsal. Some golfers believe they can improve their game by imagining a perfect swing and many gymnasts, competition divers, dancers and others try to imagine every detail of a perfect performance before they begin.

Some athletes perform specific routines under controlled conditions and their moves are always in the same order but, even in situations where we have to remain flexible, programmed and rehearsed behavior can help us perform better. Many moves in judo, fencing and other martial arts are learned and practiced as specific sequences, but, most times, an athlete entering competition has no idea which moves he will use in which order.

Instead he will watch his opponent and, as the match develops, decide that this or that sequence might be successful and he will try it. His opponent, on the other hand, may have planned to lure him into trying just that and, even after he starts one sequence, a combatant must be ready to change to a completely different one at very short notice. Again, the different sequence — and even the switch from one to the other — have probably been rehearsed.

Chess is a game for 'thinkers,' but the best players have studied games of the past and memorized strategies and sequences that are often named for great players. They may change to a different sequence at short notice, but many of their games are still based on learned sequences.

We may or may not think about our habits and procedures when we adopt them but, in most cases, we will not think when we use them. In fact many of us go through much of our daily life without actually thinking about what we are doing.

Employees in some fast-food restaurants and other businesses are guided by rules and 'procedures' that tell them how to do their jobs, how to deal with customers and even what to say. We have all met corporate or civil functionaries who cannot or will not think beyond the rules of their jobs and most of us have, in our own lives, acted from habit in a situation in which our habitual response was not optimal. Most of us also have habits and points of view left over from childhood which we know are not useful but which are very hard to shake.

We are rational and we can think things through but in many situations we rely on habit or instinct. Because we live in groups, we are also influenced by what other people do, and when one person ignores convention, the rest of the group may not approve.[98]

I'm sure that many politicians and civil servants think they actually manage the countries they serve, and it may reassure most of us to think that humans are in charge — but if we really want to control our destiny I suggest that we must first learn to understand and control both systems and metasystems, rather than allow them to control us.

A System as a Life Form

When we consider a system as a life form, its original purpose is not relevant. It may have been formed to support a prophet or to help the needy, to govern a town or to manufacture refrigerators, but when it develops into a system the original end becomes secondary. The primary purpose of a system or metasystem is to survive and grow.

While it might seem strange to compare — for example — a charity formed to protect orphans to a pirate crew that attacks and robs ships, the comparison is valid because we are talking about the form, rather than the function.

A man may work as a butcher, baker or candlestick maker but, whatever his occupation, he is still a man. In the same way a system may perform this or that function without affecting the fact that it is a system.

Some systems perform many functions, sometimes through ownership or control of other systems. National governments own and operate railways, airlines and other commercial corporations. Churches own farms, businesses and commercial real estate and the U.S. Army Corps of Engineers builds and maintains dams and waterways around the United States. In China the People's Liberation Army operates hotels and other commercial businesses.

The important question about a system is whether it has or has not reached the point at which the welfare of the system is its primary objective. If it has; then any other function, purpose or standard can and will be adjusted to promote the survival and growth of the system.

[98] Schachter, Stanley, "Deviation, rejection and communication," *Journal of Abnormal Social Psychology*, 46, 190-207)

The Village

Conventional wisdom holds that it was the invention of agriculture that began the development of civilization but author and philosopher Jane Jacobs argued that people settled in villages before they began farming. This seems to confuse many professional archaeologists because they insist that a *city* is a large settlement with big buildings, cultural establishments and a formal government. Jacobs, on the other hand, defined a city as a settlement based on industry or trade as distinct from a village — which is a service center for farms.

I respect Jacobs' definition but, because most people think of a 'city' as a large settlement, I use the term *craft village* to describe the first settlements. In fact we know that settlement began before farming because archaeologists have found and studied the sites of villages that were established hundreds of years before the development of farming in their neighborhood. Some, in Moravia and on the East European Plain, date from about 25000 BC.[99] The famous walled city of Jericho was first settled by hunting and gathering Natufians about 11,000 years ago[100] and Abu Hureyra, on the south side of the Euphrates Valley of northern Syria, was more or less continuously occupied from 13,000 to 6,000 years ago — before, during and after the beginning of agriculture in the region.[101]

Jacobs rationalized the development of the settlement she called "New Obsidian" from archaeologists' descriptions of the real-life ancient Anatolian village now called Catal Huyuk, which seems to have been a center for stone knappers[102] who had to import their raw materials.[103]

She suggested that some of our ancestors raised tool-making to a new level, even to the point at which they began to understand the internal structure of the materials they were using.

In the core-flake technique of knapping, for example, a core of stone is shaped just so, and a knapper who knows where and how to strike it can produce a finished blade with each blow of the stone that he uses for a hammer.

Early artisans learned to choose the best wood for bows and to 'season' it to make it better. Some of the goods produced by people we consider 'primitive' were, in fact, the product of sophisticated workmanship.

The Mongolian recurved bow is one example. More powerful than the English longbow and short enough to be used on horseback, it

[99] https://en.wikipedia.org/wiki/Sedentism

[100] https://en.wikipedia.org/wiki/Jericho#Natufian_hunter-gatherers.2C_c._10.2C000_BCE

[101] https://www.thoughtco.com/abu-hureyra-syria-170017

[102] "Knapping" is a conventional term for the process of chipping stone to make a tool. In some techniques a piece of a deer's antler is used to press or push chips from a stone.

[103] Jacobs, Jane, *The Economy of Cities*, Random House, NY, 1969, pg. 18-19.

was the primary weapon of the hordes that conquered much of Asia and part of Europe.

Far from simple, the Mongol bow is a laminate of wood (often birch) with layers of bone or horn, sinew and birch bark, usually held together by fish glue. Including the time for the assembly to 'season,' it took about a year to make[104] and, while most Mongols were nomadic, it seems reasonable to suspect that the bow-makers didn't move as often as their customers.

The recurve shape is also an important discovery. While most bows are a straight pole bent into an arc by the string, the Mongol bow starts out as an arc and is strung to reverse the arc and produce the recurve shape which, for any given 'weight' of 'draw,' drives an arrow much faster. Some modern 'high tech' bows use pulleys to achieve the same effect. Mongolian bows had a 'draw' — the pull required to flex it for shooting — of up to 160 pounds — about twice that of an English longbow.[105]

Even the materials used to make so-called 'primitive' weapons might be the product of a sophisticated manufacturing process. The fish glue used in Mongolian bows, for example, is made from the swim bladders of freshwater fish which are soaked in hot water to extract the protein and the water is then boiled down to make the glue.[106] The same glue can be used to stiffen and/or waterproof leather, or to stabilize birch bark.

The birch-bark resin or tar first used by Neanderthals in northern Europe to fix stone heads on spears at least 80,000 years ago is said to have been the most popular high-strength glue used in the Roman Empire.

It's simple to make — just heat birch bark and the tar drips out — but the bark will burn catch fire at a lower temperature than the tar liquefies, so you have to heat it in an airtight oven with a drain to let the liquid tar out. We can imagine Neanderthals with the skill to make and use an airtight oven — but how did they figure out the process?[107]

We might wonder how a wandering hunter or gatherer could develop a skill to a level where other members of his band would be willing to support him in exchange for goods he could make but in fact we have a modern example. Anthropologist Marshall Sahlins cites a study of two groups of Aboriginals in northern Australia in 1948, one of which maintained a full time craftsman who made tools and weapons but did no hunting or gathering himself.[108] The group consisted of only six men and three women, but they could afford to

---

[104] http://www.coldsiberia.org/monbow.htm
[105] Ibid.
[106] Ibid.
[107] http://www.primitiveways.com/birch_bark_tar.html; see also http://www.instructables.com/id/Birch-Tar-Resin-For-Paleo-Crafts-And-Tool-Making/
[108] Sahlins, Marshall, *Stone Age Economics*, Aldine-Atherton, Chicago, 1972, pp. 14-20.

support an artisan because they had to work only three to five hours a day to collect all the food they could eat.[109] Even so they did not work continuously; they would stop to rest, or for a diversion whenever they felt like it.

A nomadic life style may be attractive to hunter-gatherers — who can make the simple tools they use in a few minutes and may leave them behind when they move camp — but artisans need raw materials — wood and perhaps other materials for bows; reeds, rushes and osier for arrow makers and basket weavers, clay for potters, good quality stone for knappers and so forth; so they might be reluctant to leave a campsite that had a fire and was close to a source of raw materials.

More, wood has to season before it can make a good bow, reeds have to be dried before they can become shafts for arrows and rushes have to dry before they can be woven into baskets; so an artisan has good reason to settle in a campsite for months or even years. If that campsite had a fire, was in regular use by others, and perhaps near a convenient river crossing or where a trail branched or where two trails crossed, so much the better.

A bow-maker might settle near a stand of ash trees, an arrow-maker or basket weaver near a reed bed, a stone-knapper near a source of useful stone and so forth. Because the area around the camp is productive there is no real need to move and if the artisans' work is good, people will travel to trade with them. If possible the artisan will settle beside a river or stream and, because rivers are often territorial boundaries, the camp will attract customers from bands on both sides of the river.

The camp would also be home to some hunters and gatherers and the husband of a basket weaver might join a hunting party with other men while the wife of an arrow-maker might gather with other wives, but the artisans depend on trade for their living. They might even trade for some of their raw materials because good quality stone is hard to find but travelers from or passing through an area with good stone might bring some with them, to trade for finished products at the artisans' camp. As trade became established, some people might specialize in bringing stone from an outcropping they know and might control to trade for finished tools or weapons.

Artisans who stay in the same camp for weeks or months also have time to make better tools — like a drying rack for reeds and rushes or an airtight oven to make birch tar, for example — for their own use, and they have the opportunity and the motive to divide the labor among several people.

Humans already had a basic division of labor — men hunted and women gathered — probably because a woman in the last months of pregnancy or while nursing can still gather vegetable food but she is at a severe disadvantage for hunting. This division of labor might have led to others if we assume that male artisans were more likely

[109] Ibid. The size of the group is reported in a footnote starting on pg. 13.

to develop tools and weapons for hunting, and women more likely to develop baskets for gathering and pots for cooking. We can also guess that children might have helped or been sent to collect raw materials.

Eventually the camp will morph into a craft village and, as a 'defensible nest' it is a step toward eusociality. It still looks like a camp but it may have more residents than visitors and, as word gets out that several skilled artisans live there, hunters and gatherers might come to trade food, hides and perhaps raw materials for finished goods. Besides, artisans might like to associate with other artisans, and a skilled artisan might prefer to live in a village of artisans than travel with hunters and gatherers.

The evolution of a popular camp into a craft village was itself a big step toward a eusocial system because it increased the division of labor.

A full-time artisan in a band of hunters and gatherers was an anomaly — a person who did not hunt or gather for personal use and who made goods to be used by others. The craft village was based on artisans who made goods for people who were not members of their family and who ate food that was hunted or gathered by someone outside their immediate family.

At this stage probably all the villagers were competent hunters or gatherers, but as new generations were born, some did not learn to hunt or gather as well as others and some did not learn the skills to make high-quality tools, weapons, baskets or other goods. As each individual learned more about one specialty, he or she learned less about others. To cite a modern adage, 'we learn more and more about less and less.'

Swarm Intelligence

Like a camp, the craft village was probably ruled by the kind of swarm intelligence that seems to operate in ant hills, bee hives and hunting and gathering bands.

Among ants and bees the queen is supreme but, as far as we know, she does not make decisions. She is tended and served, but she is an egg-laying machine that must be tended, rather than a ruler who must be obeyed.

Individual insects within the colony seem to be ruled by their perception of what is needed, and the colony as a whole seems to be ruled by consensus.

Among eusocial bees, jobs are determined largely by age. Young workers build and/or maintain the hive and tend larvae and older bees are foragers. Some foragers collect nectar and some collect pollen, but they can shift from job to job.[110]

[110] Michener, Charles D., *The Social Behavior of the Bees*, Belknap Press of the Harvard University Press, Cambridge, Mass, 1974, pg. 64.

But even as they do different jobs, insects in a eusocial colony appear to be equals. In most cases the queen decides where the nest will be but that's because she digs or builds the nest herself, before there are any workers to help her. Once the nest is established workers can make decisions and, in an experiment, worker bees chose the site for a new hive.[111]

When a colony of bees gets too big or a disaster drives them from their hive, bees will swarm and look for a new site. In an experiment, Cornell University biologist Thomas Seeley, Kirk Visscher of the University of California at Riverside, and others moved several swarms of bees to tiny Appledore Island off the coast of Maine, where the University of New Hampshire maintains the Shoals Marine Laboratory. Scouts from each swarm searched the island and returned to their swarms to report their finds with a waggle dance similar to the one workers usually use to report a good source of nectar. In response to the dance, other scouts flew to inspect the recommended site and when enough scouts — usually about fifteen — approved it, the swarm moved to it.[112]

Studying red harvester ants in the Arizona desert biologist Deborah Gordon found that a few 'patroller' ants leave the nest in the early morning, apparently to check the weather and other conditions (such as the presence of predators) while most of the workers wait inside. It seems the patrollers smell different from the foragers and when the patrollers return to the nest, they touch antennae briefly with the foragers.

If the patrollers don't return there may be danger outside and the foragers don't go out. If two patrollers return and touch antennae with foragers within ten seconds of one another, the foragers start to go out. This may be based on scent because in an experiment Gordon and her collaborator Michael Greene captured the patrollers as they left the nest then, a half hour later, dropped glass beads with patroller scent into the nest entrance. When they had dropped enough beads, the foragers began to come out.[113]

Most human colonies or groups have leaders, and that's both a blessing and a curse. We don't know how that happened, but I can speculate. That may not sound very 'scientific' but let's remember that anything we say about very early prehistory has to be speculation, and at least I admit it.

Some members of a hunting and gathering community might be more respected than others and special skills would be recognized, but essentially each individual chose his or her own life style and made his or her own decisions.

[111] Miller, Peter, "The Genius of Swarms," *National Geographic* magazine, July 2007, pg. 130-147.
[112] Ibid.
[113] Gordon describes her studies in *Ants at Work*, published under The Free Press imprint by Simon and Schuster, NY, 1999.

That sounds like heresy to many people. How can a group of people live together if they all 'do their own thing' and there is nobody to rule or co-ordinate? Sounds unlikely, but it seems to work.

In the same way, hunting and gathering bands seem to operate well with no formal organization. When a man wants to hunt he may join a group that is starting out or, if none is starting, he might start one. When a woman wants to gather she might join a group or start one. Some hunters are better than others and more people would follow the best hunters, but they follow because they chose the leader, not because the leader chose them.

And the same leader might not lead all the way. Among hunters one man might be known to be best at finding game, another best at stalking and yet another the best shot with bow and arrow or able to throw a spear farther and/or more accurately than others. Each one would take — or rather be given — the lead at appropriate times.

I've seen such a shift in leadership on a trip from the village of Rankin Inlet, on Hudson's Bay in Canada's Nunavut Territory, about 500 miles south to the town of Churchill in the province of Manitoba. I travelled with four Inuit in two twelve-seater Bombardier snowmobiles and, as we travelled, the leadership shifted back and forth between two very capable elders.

Yvo Airut has a freight-hauling and construction company and a garage where he repairs his own and others' trucks and heavy equipment. Joe Kaludjak is Yvo's cousin and hunting buddy and was, at the time, part owner of a freight boat and president of a corporation that invests the band's funds.

On land, Joe took the lead. He chose the route and, because his snowmobile was faster than Yvo's, he sometimes got so far ahead that we lost sight of him — but when he came to a big river or to a stretch of sea ice across an inlet, he would stop and wait, sometimes for an hour or more, for Yvo to lead the way. Joe knew the route and Yvo was happy to let him lead, but both knew that Yvo was better at judging ice. Most arctic ice is safe, but some river crossings are dangerous because ice can erode from below or can be very thin where a sand bar has washed out after the ice formed. At one point when we were crossing a bay on sea ice Yvo stopped, went ahead on foot and tapped the ice with a steel bar, then led us back to land and a long detour.

Both Joe and Yvo were respected for their skills, but they knew and respected each other's skills and they traded the leadership back and forth as appropriate.[114]

Among women, one might know better routes or be better able to predict where specific berries or fruit would be most plentiful or at their best, and others would follow her. Another might choose,

[114] I wrote about this trip for several magazines and one newspaper. The newspaper article is on line at http://www.japantimes.co.jp/life/2000/06/14/travel/bombardiers-and-polar-bears/#.VieiGCihifQ

for one reason or another, to go by a different route or to a different location, and some might follow her.

Like members of a hunting and gathering band, the inhabitants of a craft village are individuals who live together because they are friends and, while a band may control the territory, all members of the band have equal rights.

Trade

The craft village brought the division of labor to new heights for a human community. Here, for the first time, many humans depended on others, outside their immediate families, to provide for their needs.

For millennia some humans had hunted and some had gathered, but, in most cases, the hunters were mated to the gatherers. Even so there was some sharing, because most hunters share their kill with everyone resident in the camp, but the village brought this to a new height.

There were probably both hunters and gatherers living in the village but the artisans didn't hunt or gather and, for the most part, the hunters did not make their own weapons or nets and the gatherers may not have made their own baskets. The weapons, nets, baskets and other equipment were made by specialists who did not hunt or gather themselves and who may have had help to produce them. In fact we see here an early version of the economic division of labor that I call Ricardo's rationale.

David Ricardo was an early 19th century English economist who suggested that communities and whole countries should specialize and produce only the goods they could produce more efficiently than any other. Even when two countries could produce the same kind of goods, he said, each one should produce the kind of goods in which it has a "comparative advantage."

He wrote,

> Two men can make both shoes and hats,...and one is superior to the other in both employments; but in making hats he can only exceed his competitor by one fifth or 20 percent, and in making shoes he can exceed him by one third or 33 percent. Will it not be in the interest of both that the superior man should employ himself exclusively in making shoes and the inferior man in making hats?
>
> Thus even if one community can make every product more efficiently than another, it should specialize only on those items it produces most efficiently, in relative terms, and trade for others. Each community, and ultimately each nation, should specialize in what it does best.[115]

[115] quoted in *The Case Against the Global Economy*, pg. 220, in David Morris' article "Free Trade, the Great Destroyer."

Taken to the extreme, this is the rationale for the modern 'global economy' and the problems it causes, but on a local level it makes good sense and some artisans who did not hunt or gather their own food might not even collect their own materials or make all their own products.

A skilled basket weaver could make better use of her time weaving baskets than collecting willow withes, and she might delegate an apprentice who wanted to learn the trade — or perhaps someone who never planned to weave baskets herself but wanted to earn one — to collect them for her. A skilled arrow maker might not collect his own reeds, and a skilled potter might employ help to dig and carry clay.

An arrow maker might send his wife and/or children to gather reeds and trade with a stone-worker for the arrow heads and perhaps with a hunter for the feathers. A potter probably would not need his or her kiln every day and it might make sense to share with a neighbor. Because a kiln has to be watched while pots are being fired, it might make sense for one artisan to build and tend a big kiln and fire pots for several potters. We can only guess what cooperations might have developed, but it's a safe bet that some did.

As more and more artisans settle in the village it changes its nature. Jacobs would call it a 'city,' but I prefer the term craft village because most people think of a city as a big settlement, whereas the population of a craft village might be only a dozen families or less.

As Jared Diamond notes in *The World Until Yesterday*, strangers may not be allowed to travel in a band's territory because a stranger might be a scout, planning a raid or planning to steal something or perhaps to kidnap one of the women. Either way, the safest course is to kill him.[116]

But this is not an absolute prohibition. Hunting and gathering people tend to be self-organized on two levels — in bands that seldom have more than 25 or 30 members and in larger groups — which we might call mating networks — which might include ten or twenty bands. This dual organization is necessary because a large group is inefficient for hunting and gathering but, for the most part, young people raised together in a small group won't mate. The mating networks are typically loosely defined, and a band that lives close to the border between two networks might be a member of both of them.

Bands that are members of a mating network often party together and some members of most bands have been adopted into the band by marriage, so a camp or village that is not open to strangers might well be open to members of other bands within the mating network.

If people from distant bands could not travel to the village, goods from the village could still travel as gifts. Neighboring hunting and gathering bands often exchange gifts and if one has an outcropping of obsidian in its territory then a chip or chunk of it would have made a valued gift. Obsidian makes wonderful cutting tools — chips of it are

[116] Diamond, Jared, *The World Until Yesterday*, Penguin, NY, 2012, pg. 50.

so sharp they make cuts that heal without scars and modern cosmetic surgeons use them, in steel holders, as scalpels. People who receive a chunk of obsidian would reciprocate with the most desirable gifts they could offer — choice fish, game or fruits, pretty feathers or whatever — so the owners of the outcropping of obsidian would be encouraged to give more and more of it as gifts.

Suppose band A has a source of obsidian, and gives gifts of it to band B. Members of band B also exchange gifts with band C and, when they have enough obsidian for their own use, what better gift to pass on? Eventually members of band C will start giving gifts of obsidian to band D, and we have the start of a network.

In time, travelers might be allowed to pass through surrounding territories on their way to and from the village. In New Guinea, Diamond notes, members of one band were allowed to pass through another band's land to reach traders who lived by the sea, but not allowed to hunt or gather on the way.[117]

Now suppose someone from Band A — which has obsidian — realizes that some of the baskets he gets in exchange for obsidian are exceptionally well made. He knows they don't come from band B, and he asks about them.

Members of band B may be reluctant to tell him, at first, but he has this wonderful obsidian and he offers a gift of it in return for information. All his band B informant can tell him is that they get them from band C but, in exchange for obsidian, he may agree to escort the would-be trader band A across band B's territory to meet with someone from band C.

The trader would have to pay his escort and then hire another to cross the territory of the next band, but if he has obsidian — a valuable commodity — he could afford it and when he finds the source of the baskets he would find that, if he pays in obsidian, they are very cheap. He may trade for other things along the way — always on a gift-for-gift exchange — and eventually he will be allowed to travel long distances because he has friends, suppliers and customers in every band whose territory he passes through.

Trade was very profitable for the traders and good for the people they traded with and, even in the Stone Age, trading networks spread far and wide. English archaeologists discovered in the late 1940s that many of the stone axes they found at ancient sites over most of England came from three sources: a mountain in North Wales, Rathlin Island in Ulster and a site near Great Langdale in the Lake District. Near the quarries, archaeologists found debris from the making of thousands of axes.

Axes made of jadeite stone from the Swiss Alps were used in Brittany, Germany, Belgium, England and Ireland and axes made of obsidian from one mountain in Turkey are found around the Mediterranean and most of the Middle East.

[117] Ibid., pg. 40.

Neolithic farmers around the present site of Oslo, in Norway, used axes made of flint from north Jutland, and new and apparently unused adzes from southern Sweden have been found by the Gulf of Bothnia, a northern bay of the Baltic Sea, about 900 miles from where they were made.[118]

Smaller blades such as knives, scrapers and arrowheads made of flint were traded too. A flint mine at Grimes Graves in Norfolk, about 80 miles north of London, England, eventually covered about 35 acres and had more than 360 shafts, some of them 40 feet deep. It was in regular production for about 600 years, around 2,300 BC, and archaeologists say miners at the site used more than 50,000 red deer antlers as picks.

A series of ancient flint mines at Krzemionki in the Opatow district of Central Poland seems to have had about 1,000 shafts over an area of 800 acres. Each shaft was about six meters deep, with several galleries radiating from the bottom of each one.

The crowning glory of the era was the ground-and-polished Neolithic axe, so important that it is seen as a marker for an era of human pre-history.

Flint axes dull quickly and may shatter in use. The Neolithic axe is made of jade or obsidian, both of which are tough as well as hard. Like other early tools the Neolithic axe was first chipped to a rough shape but, unlike earlier tools, it was then ground and polished to a fine finish. It was tough enough for heavy work, it took and kept a good edge and, in some ways, it was better than modern steel axes. *Encyclopedia Britannica* says that in one experiment Danish scientists took some 4,000-year-old polished-stone axes from a museum and fitted them with new handles, copied from an axe found in a bog with the handle preserved. The Neolithic axe-handle was shorter than a modern handle and it takes a different technique — but once you learn the technique it is very effective.

With a bit of practice the Danish experimenters could fell an oak tree more than a foot in diameter in an hour or a two-foot pine in two hours. Three men cleared 600 square yards of silver birch forest — enough for a small subsistence farm — in four hours. The axes had not been sharpened for 4,000 years but they were still in good shape after cutting down more than 100 trees.[119]

With Neolithic axes men could fell trees to make log houses or to cut roof-beams for stone or brick houses, build stockades and bridges, split cedar into shingles or logs into planks and build big buildings with clear-span roofs supported by wooden beams. By mounting a similar (or perhaps the same) head sideways on the handle, they can make an adze — which can be used to hollow out logs to make dugout

[118] Clarke, Grahame, *Economic Prehistory*, Cambridge University Press, Cambridge, 1988, pg. 169-198.
[119] *Encyclopedia Britannica*, 1998 edition, 28:696, early axes, 28:442.

boats. With the development of polished-stone tools, men developed building techniques that we still use today.

One stone-axe factory kept going until about 1830 at Mount William, near Melbourne, Australia, in territory controlled by the Wurrunjerri tribe of aborigines. The last operator was a man named Billi-Billeri, who died in 1846. He sold blanks — rough-shaped axes, the prehistoric equivalent of unpainted furniture — which his customers finished by polishing and sharpening. In the final years only he and his sister's son mined the stone and roughed out the axes, but in earlier times there may have been more people, because axes from Mount William were traded among tribes that lived more than 100 miles away.[120]

Axes may also have been a medium of exchange. Unfinished axes have been found among the grave goods of stone-age burials and later, in the Bronze Age, ingots of copper from mines near the Rhine and Weser were often cast in the shape of a double-bitted axe.

Trade routes and networks developed around the world. Anthropologist Marshall Sahlins describes one South Pacific network in *Stone Age Economics*.[121] Siassi Islanders live on and around a group islands in Vitiaz Straits between the islands of New Britain and Umboi, near New Guinea, with so little land of their own that many of them live on houses supported on stilts above shallow lagoons. They occupy about one third of one percent of the land in Papua New Guinea's Umboi governmental district but they make up about one quarter of the population and they are among the wealthiest people in the area.

They do some fishing but they live mostly by trade. The area's only source of obsidian is on a peninsula of New Britain, more than 100 miles east of the Siassi Islands and the best potters in the area live in New Guinea, about 50 miles to the west. The islanders also deal in pigs from New Britain and sago from Umboi and, to a lesser (but still very profitable) extent, luxury goods such as boars' tusks, dogs' teeth and wooden bowls. Young men in New Guinea, Umboi and New Britain have to buy such goods as gifts to make a wedding proposal acceptable.

To protect their source of supply, traders may be vague about where the goods they offer come from or how they are made. Clay cooking pots, for example, were made at three locations in New Guinea but none in Umboi or New Britain, and some of the people who bought them were apparently told that they were the shells of deep-sea mussels.[122] Among Western cultures, the ancient Romans thought silk grew on trees and most Europeans of the era thought steel came from secret Asian mines.

[120] Clarke, Grahame, *Economic Prehistory*, pg. 182-3.

[121] Sahlins, Marshall, *Stone Age Economics*, Aldine-Atherton, Chicago, 1972, (reprinted 1992, 2005), pp 282-284.

[122] Ibid.

We don't know how far the Siassi islanders' network extended but author Jared Diamond reports that obsidian axes from quarries on New Britain are found from Fiji, about 2,000 miles southeast of New Britain, to Borneo, about 2,000 miles to the west.[123]

The most famous trade route of the ancient world was the 'Silk Road' that stretched about 4,000 miles from the city of Sian, on the border of ancient China, across central Asia to the Eastern Mediterranean where European traders loaded the goods on ships for delivery to Greece and Rome.

Along with silk, traders from the east brought roses, azaleas, chrysanthemums, peonies, camellias, oranges, pears, the crossbow, gunpowder, printing and paper to Europe. In return they carried watermelon, grapes, grape wine, alfalfa, figs, pomegranates, cucumbers, sesame, chives, coriander, safflowers, horses and Bactrian camels to China.

Damascus, Baghdad, Tashkent, Samarkand and a dozen other cities began as caravan stops on the silk road, and they became great cities of their day because the products and the ideas of the world passed through them. Chinese Jews and Nestorian Christians got their religions over the silk road and Buddhism came to China from India over a branch of the silk road.

By Roman times, the trade was well established and the Romans spread their own trade networks from the silk road throughout Europe. They built some of the best roads in the world to move their armies on, but those roads also supported the trade that paid for the armies.

The Silk Road continued through the fall of Rome but the western end of the route was nearly abandoned for several hundred years after the wars that spread the Muslim faith across the middle east made travel through the area unsafe. It was re-opened about AD 1200 when the Mongol empire of Genghis Khan brought peace to the area.

Venetian trader and adventurer Marco Polo gets much of the credit for opening the route from the European end but only because he was the first to write about it. In fact some scholars doubt that Polo ever went to China because he never mentions tea, which was a major feature of Chinese culture and one that was completely unknown to the Europe of his time. Everything he did write about was already known in Europe, and there were some Europeans in Khan's court. Few people travelled the whole length of the route, but while Khan lived, it was a major trade route and goods and even mail were relayed back and forth along it. One caravan of more than 200 camels made a round trip every two years to bring tea from China to Catherine the Great in Moscow.

The Silk Road was in regular use until the opening of the Trans-Siberian Railway in 1908. And the route may be coming back for another round. In the 1990s, I travelled much of the northern route

[123] Diamond, Jared, *The World Until Yesterday*, Penguin, NY, 2012, pg. 72.

of the Silk Road — through Moscow — with truckers who delivered a main-frame computer from England to the head office of the Bank of Uzbekistan in Tashkent.[124] Some of the truckers I rode with had hauled goods from England to Alma Ata, Kazakhstan, near the Chinese border, but I'm told that English truckers have now abandoned that run because there's too much competition and prices are low.

Trade routes within the Americas have not been well mapped, but we know that the Inca, based in Peru, traded over more than a thousand miles up and down the west coast of South America. Some evidence suggests contact between the Inca and the Aztec, who lived on the present site of Mexico City, and we know that the people of Cahokia, near the modern city of St. Louis, traded with the Aztec as well as with the Cree, who lived north of the Great Lakes, and east and west to both coasts of North America.[125]

Routes within Africa are not as well-known but the ruins of the city known as Great Zimbabwe date back to about AD 800 and the site was occupied long before that.[126] The Karanga people who built the city appear to have traded with India, Arabia and southeast Asia. Most of this trade seems to have been carried in Chinese ships, but Arabs had ships that could have traded over the same routes.

There is also evidence of travel and trade between the Americas and West Africa long before Columbus 'discovered' America. The natives Columbus met on Hispaniola told him of black traders who had spears with tips made of a gold alloy, similar to some used in Africa at the time, and on his third voyage natives on the coast had cotton kerchiefs with patterns similar to those used in Sierra Leone and Guinea.[127]

If that sounds unlikely, consider that German psychologist Dr. Hannes Lindemann crossed the Atlantic alone in an African dugout canoe in 1955. Dr. Lindemann is better known for his later crossing in a mass-produced German kayak, but most descriptions of the crossing by kayak include references to his crossing by dugout.[128]

There may also have been some early trade across the Pacific. As a student at the University of Pennsylvania Dr. Peter Harrison found some Chinese coins dating from about the time of Christ when he excavated the central acropolis of the Mayan city of Tikal.[129] Clay pots, similar to those produced by the Jomon culture of southeastern Kyushu, thousands of years before Christ, have been found in Ecuador.

[124] This trip is described in my book *Road to Tashkent*, available as an ebook on Amazon. See also *Japan Times*, March 27 1996, p 15.

[125] Information on the trade of Cahokia is from pamphlets distributed at Cahokia Mounds, World Heritage Site and State Historic Site, Collinsville, Ill.

[126] Garlake, P. S. *Ruins of Zimbabwe*, Lusaka National Educational Co, 1974.

[127] Van Sertima, Ivan, *They Came Before Columbus*, Random House, New York, 1976, pg.11, 14.

[128] His crossing by kayak was the cover story of *Life* magazine, July 22/57.

[129] Dr. Harrison told me about the coins in a personal conversation, shortly after he completed his work at Tikal.

Thousands of glass fishing floats, of the type used by Japanese fishermen, and debris from the Japanese tsunami of 2011, are found along the west coast of the Americas and dozens of disabled Japanese fishing boats have drifted across the Pacific. Many of the fishermen in those boats died but some survived, and when Norwegian explorer and writer Thor Heyerdahl sailed a balsa-wood raft across the Pacific Ocean from South America, he and his crew were able to catch fish to eat and rainwater to drink all the way across. Some west coast Native North American languages include words that may have originated in Japan.

In February of 2014, a lone fisherman from southern Mexico landed on one of the Marshall Islands. Jose Salvador Alvaredo explained that he lost his bearings in a storm, and he lived on raw fish, birds and sea turtles for about 13 months as he drifted across the Pacific Ocean.[130]

Trade was vital to the development of civilization because the traders who carried goods also carried ideas. That was important because, in any closed society, ideas tend to crystallize and misconceptions are seldom challenged. People who live in isolation have little chance to develop because they get few new ideas from the outside and have no way to pass their own ideas to anyone else. If someone invents or develops a new way of making fire, thatching a hut, tanning leather or whatever it will spread through the band, but then what? They may party or fight with neighboring bands, but fights and parties are not good occasions on which to discuss ideas.

But traders talk. They may be party people at heart and they are also gossips. The first trade was incidental to social events and, even now, a good salesman will try to develop a social connection before he tries to make a sale.

So traders need something to talk about when they arrive in a new village. If they have seen or heard of a new way of making or doing something, that's a logical topic of conversation.

If bands live an average of ten miles apart and party together three times a year, then even if they talk about it a new idea will take more than thirty years to travel a thousand miles. If three groups of traders cover an average of 400 miles each, the same idea could travel that far in a year.

Even the goods themselves contain ideas. The trader who brings a stone axe, for example, also brings the idea that a certain kind of stone makes good axes. In some cases the band that buys the axe may have that kind of stone in their territory, but they may never have tried making an axe of it.

The way the head is mounted and even the kind of knots that hold it in place may also be useful information.

[130]http://www.telegraph.co.uk/news/worldnews/australiaandthepacific/marshallislands/10616194/Castaway-from-Mexico-First-photos-of-Jose-Salvador-Alvarengas-boat.html

One way or another, the development of trade seems to have paralleled the spread and even the development of ideas. Archaeologist Brian Hayden lists some of the developments of the Mesolithic period in *Archeology, The Science of Once and Future Things*.[131]

Awls made of bone enabled Mesolithic people to pierce holes in leather and use sinew to sew pieces of leather together to make tents or clothes. Baskets woven of willow twigs made it practical to collect berries and bring them back to camp. If the baskets were sealed with fire resistant clay, they could be used to carry water and even for cooking and, in time, these sealed baskets probably evolved into pottery.

Baskets also open up new food sources because while some nuts — acorns, for example — are toxic, they can be eaten if they are ground with rocks and the meal soaked in water to wash out the bitter tannin compounds. Ancient Greeks, Japanese and Iberians ate acorns and, in historic times, they were a staple for some Native North Americans in California.

People also began to eat fish in the Mesolithic. They probably ate them in the Paleolithic too, because you can catch fish with your bare hands, but they ate more after they invented fish-hooks, pronged fish spears, nets (also useful for catching animals) and several types of fish traps.

Mesolithic hunters invented the bow and arrow — a quantum leap in weapons — and they domesticated dogs. They invented sleds and canoes for transport and learned to store food in pits, and to store or dry it on scaffolds.

Along with the development of trade, our ancestors developed the concept of treasure. A ruby is just a red pebble, but after a craftsman has cut and polished it and mounted it in a ring, bracelet or tiara, it's a thing of beauty. Even in its raw form, it may be something to trade.

Most Paleolithic tools could be made in a couple of minutes and they might be thrown away after one use, but some goods took hours — perhaps days — to find the right materials and to make. Archaeologists estimate that a flute made from a bird bone found in one German cave probably took only a couple of minutes to make, but an ivory flute found in the same cave probably represented more than 100 hours of skilled work.[132]

Amber from around the Baltic Sea has been found in caves in Crete, and Neolithic beads, bracelets and pendants made of the shells of mussels found only in the Black and Aegean Seas are found as far north as Poland. One lot of 20 bracelets found in eastern Bavaria

---

[131] Hayden, Brian, *Archeology, the science of once and future things*, W H Freeman and Co, NY, 1993, pg.6.
[132] www.bbc.co.uk/news/science-environment-18196349

suggests that the shell was probably worked at the source and traded as finished jewelry.[133]

The Agricultural Revolution

It may have been in the fields around the village that so-called 'domesticated' grains evolved. Many archaeologists believe they were deliberately cultivated, but I find it easier to believe that they evolved naturally.

The craft village didn't need farms because the land around a hunting and gathering camp or a craft village would have been more productive than land farther away from it.

Besides, hunters and gatherers who came to trade with the artisans would have brought food to trade for the artisans' products. Because they brought fruits, vegetables and grains from all directions the garbage middens near camps grew a wide variety of plants, and just about every cross possible would occur.

One difference between the wild Einkorn wheat that Jack Harlan collected and modern domesticated wheat is that while Einkorn has one seed per plant, domesticated wheat has many. We eat the seeds of grain, not the stalk, and people who collect wild grains would naturally choose varieties that have more seeds. If they bring the grain they collect to a village to trade it then, by accident, some of the seeds might be spilled or dropped near the village. If they grow in the same field, it's no surprise that varieties that produce many seeds might displace varieties that produce only one seed per plant. Einkorn might survive — as it does — in areas where there is little or no competition from varieties that produce many seeds but if gatherers bring all varieties of a plant to a village, the competition around the village will be intense.

Another characteristic of domesticated grains is that they don't shatter. Wild grains drop their seeds when they are ripe but the seeds of domesticated grains stay on the stalk until they are threshed. Conventional wisdom tells us that this characteristic was deliberately bred into grains for the convenience of farmers but I find that hard to swallow.

Whether you like the domesticated or evolved scenario you must, unless you want to postulate genetic manipulation, assume that some wild grains did not shatter. In the domesticated scenario these might soon die out of a cultivated field because most grains would drop their seeds in the field while the seeds of grains that did not shatter were collected and eaten. In the evolution scenario the grains that did not shatter were collected and taken to the village to trade. Some of their seeds would be lost in transit and, because they were all coming

[133] Prof Grahame Clark outlined the trade patterns of prehistoric Europe in Chapter IX of *Prehistoric Europe*, Methuen & Co Ltd. London, 1952 (1965).

to the village, there would be more lost in the fields around the village than anywhere else.

As more and better grain grows around the village grain-eating animals would move into the area and, even if some are killed, animals that live near a village will eventually lose their fear of man. Some might even live in the village, because people sometimes adopt young animals as pets.

This is very much like farming, but the people of a craft village probably did not take the final steps to land ownership and cultivation, because they didn't have to and because grain farming offers questionable benefits to the farmer. In a good year a farmer can produce more food than a hunter or gatherer but he has to work much harder to do it and, at this point, he has no use for the difference.

It's hard work to farm and use grain but that's not a problem if you have slaves. Through the ages most farms have been worked by slaves and in some times and places — such as the Roman latifundia, the sugar plantations of colonial Jamaica and the cotton plantations of the American south — men and women were bred and worked like animals.

The conventional view of history sees this as a positive development but this view was developed by rulers and priests, who benefitted from the farms, rather by the slaves who had to do the farming

I suggest that the so-called 'invention of agriculture' was actually the invention of large-scale long-term slavery. Through most of history farms have been worked by slaves or by serfs who are, in effect, worse off than most slaves. The difference is that slaves are an investment and if the crops fail slaves have to be fed anyway. Serfs are responsible for their own maintenance, and they can be left to starve.

Life on a farm is hard for women, even if they are not slaves. Berries, fruits and many roots can be eaten directly, or thrown into a stew together or with meat, but grain has to be mashed or milled and then mixed with water and cooked into porridge or biscuits to make useful food for humans. The grinding is a hard and slow job, and possibly painful. Archaeologist Elizabeth Barber says that toe, knee and shoulder bones of many women who lived in the early farming villages of Mesopotamia were deformed by kneeling and pushing as they ground grain with the type of stone grinders found in those villages. What passed as work for hunting and gathering women was literally a stroll in the woods but an early farmer's wife was virtually tied to a grindstone.

A farmers' diet is also a problem, because people who grow wheat eat wheat, but wheat alone is not a good diet. Hunters and gatherers eat a wide variety of food — archeologists typically find traces of 75 or more food plants in a typical hunter gatherers' campsite — but even in recent history, woods and fields near a village are 'owned' by the lord and peasants are not allowed to hunt or gather. Robin Hood was a hero because he lived in the forest and ate 'the king's deer' and even now wealthy land-owners in England and some other European countries employ gamekeepers to prevent poaching on their land.

In fact the so-called 'agricultural revolution was not the invention of agriculture. Most modern anthropologists believe that agriculture began more than 10,000 years ago when someone living near the middle of Africa discovered that if you eat only part of a yam and throw the rest away, the part you throw away will grow into a new yam. If you throw the tag ends of all the yams you eat into the same field, you have a field of yams the next year. The first domesticated food crop seems to have been the white Guinea Yam.[134]

Yams are easy to harvest and store because you don't have to. Grain has to be harvested when it's ripe and then stored somewhere but yams can be left in the ground until you need them, and yams that are not dug up and eaten over the winter will spawn a new crop next summer. As in yam-gardening cultures today the gardens may have been planted and tended by women but the work is light and the hours are short so women still have time to gather wild foods, and their men can hunt.

Yams were probably the first crop that was actually 'farmed,' but the roots of farming go back probably thousands of years beyond the first yam garden. An article in *New Scientist* magazine[135] says that modern botanists realize that people around the world have been tending fields and forests for tens of thousands of years, clearing brush around selected fruit trees to give them space to grow, setting fires to encourage new growth and generally shaping the natural landscape to include more of the plants most useful to man.

Even in historic and modern times Australian Aboriginals, who do not maintain farms, carry seeds of useful plants as they travel and plant them near good campgrounds. Like hunters and gatherers the people of the craft village probably understood agriculture but they did not need it; and they did not want it because hunting and gathering provides a better variety of food for less work. As noted earlier, Prof. Jack Harlan tells us that virtually every known group of hunter gatherers knew enough about plants that they could have farmed at any time, but they had no need to.[136]

The Conquest

I suggest that the development of agriculture was an indirect result of the development of trade, which developed the concept of wealth and of robbery.

Traveling traders might be tolerated but, at a time when it was still legitimate to kill strangers, some of them would have been killed and the killer would have taken his victim's goods. Some people might

134 *Encyclopaedia Britannica CD 98*, "Central Africa\the region\history\early society and economy\the agricultural revolution."

135 Holmes, Bob, "Quiet Revolutions," *New Scientist*, October 31, 2015, pg.31.

136 Harlan, Jack, *The Living Fields*, Cambridge University Press, 1995, pg. 15.

find robbery more profitable than hunting, gathering, working as an artisan or trading, and while a robber might not rob someone from his own band, he could waylay traders from other bands and take the goods to trade himself.

As the danger of robbery developed traders might begin to travel in caravans, for protection, and robbers might gather in gangs big enough to rob caravans. Because the robbers form gangs the caravans hire guards and because the caravans have guards the robber bands combine, and grow from bands into armies.

Professional robbers are another division of labor and so are the guards, camel-pullers and others who accompany caravans. There's a division of labor within the robber bands, too, because one man leads and others follow. The band might also keep some slaves, to build and maintain their camps and to carry their loot.

Robber bands might grow by joining forces, and they might also recruit survivors from among the surviving guards of caravans they rob. The guards of the caravan were enemies of the robbers, but many soldiers don't care who they fight for and, given a choice, many will switch sides rather than die. Most of the empires in history enlarged their armies by enlisting defeated foes but the first time it happened was a big change because, up to this point, hunting and raiding parties were groups of close friends with no outsiders.

The idea of an alliance with strangers is radical but it works. The band that accepts former enemies as recruits becomes bigger and it can raid more effectively. Because it raids effectively more recruits join it and, because it continues to accept recruits, it gets bigger and stronger than other bands. Eventually, it becomes an army.

This is a new organization because the army is very different from the kinship and friendship-based organizations that came before it. Perhaps the biggest difference is that it does not have the kind of priorities that humans are used to.

Up to this point most human groups have been based on families and the first priority of a functional family is the welfare of the children. The survival of the parents is a factor in the welfare of the children but, if necessary, many parents will die to protect their children.

We know this is normal behavior among many animals and we can assume that the protection of children is inherent in human nature. It must be, because families that did not protect their children would have died out. The families that survived to populate the world were the ones that protected their children.

In a family, the strong protect the weak. If a tiger threatens a family the father faces it first, backed up by his adult sons. In an extended family the best fighters will lead an attack or defense and all the men will protect all the women and children. If a family runs short of food or water the children, pregnant women and nursing mothers will get full rations while men go without.

An army works the other way round. The commander of an army may be the strongest warrior and he may lead his men into battle but he does not feel a duty to protect his subordinates — especially not the ones who were his foes last week. In an army the common soldiers are the first to be sacrificed, and They have no choice in this. The fighting men of a family may sacrifice themselves if necessary but the sacrifice will be voluntary and the beneficiaries will be their family. The sacrifice of soldiers in an army is at the discretion of the commanding officer and it may be for objectives that mean nothing to the soldiers.

Many of the soldiers may have no bond of kinship and the plans of the army may have no direct relationship to the welfare of the soldiers. Whatever that purpose is, soldiers are expected to place the interests of the army ahead of their own.

An army will make sure that its soldiers are well fed but it cares for them as a workman cares for his tools — so they will be in good condition when they are used.

The concern of a family or a clan for its members is different. If a member of a family is wounded, the family will take serious risks to save him. If a member of a hunting party is seriously injured, the hunt may be canceled while the others care for him.

If a soldier is wounded he may be cared for or he may not. Either way it is not likely that a war, raid or other action will not be called off while the officers tend his wounds. Even in peacetime, the welfare of a common soldier is not allowed to interfere with the plans of an army and it may not be allowed to interfere with the convenience of its officers.

We in the modern world are used to this attitude. We don't expect systems to worry about the welfare of junior members but up to this point all human organizations were based on families and, in a family, the welfare of the junior members is a primary concern.

As the band grows there is a further division of labor when the bandit chieftain appoints sub-chieftains, and even more when it becomes an army.

Robbers may attack a caravan, but armies seek bigger prey. The first army to take a village probably plundered it and moved on, but now we see another evolutionary development.

Taking the Village

After a caravan is robbed the traders have nothing left to steal. A village produces wealth and if you rob it this month, you can rob it again next month. The hitch is that a village that is robbed often will develop defenses so, rather than rob and run, it makes sense for an army that takes a village to hold it.

Now the robbers have a base, but traders can avoid them by avoiding the base. Robbery is good business, but it's bad policy to rob people on your own ground. The village is a source of wealth but it produces only a few goods and, to cash in, the robbers have to encourage trade. They can do that by taking only a portion of a trader's goods, which they can call a 'toll' or 'duty,' and by protecting traders from other robbers. In fact the robbers who waylay caravans on one route may be the soldiers who guard them on another, but the soldiers are not allowed to rob people who have paid the toll and who remain within the protected area.

But now the village must be 'protected' and the residents — especially the artisans whose products are so valuable — must not be allowed to leave. As the army settles in it will enslave some villagers and force them to build a wall around the village and a fortress for the soldiers to live in.

Because the village was probably beside a stream it did not need a well but even some animals know that if you dig a hole near a stream you will hit water, and the soldiers will demand a well within the wall. Food gatherers have to leave the village but they will work under guard, in fields around the village, and they will leave their families as hostages within the walls.

The Occupied Village

It was in the occupied village that our modern world took shape. Up to this point human communities had consisted of friends and relatives, all with equal rights, but the occupied village was now home to both the army that conquered it and the artisans who built it.

By modern standards the first army was a small band of thugs and the village they took was barely a hamlet, but the conquest and occupation of that village led to class distinctions among people, slavery and the 'battle of the sexes,' land ownership and farming, the cash economy, poverty and conspicuous consumption, social conflict, formal religion, government and, probably, a significant change in the nature of humanity. We have to deal with these developments one at a time but, as we do, please understand that they all took place more or less simultaneously.[137]

Up to this point there were probably no class distinctions in the human world. Among hunters and gatherers all are equal and — aside from the distinction that men generally hunt and women generally gather — everybody lives the same way. Some people have more prestige than others but in most cases the prestige is related to a specific skill. One man might be known to be the best tracker and another the best shot with a bow and arrow or be able to throw a spear farther and more accurately than others but, in general, all members of the group are equal. Anthropologist Paul Radin says:

[137] This is a rationalized reconstruction. For convenience I write as though everything happened in one village, but in fact we know that the social order that we call 'civilization' must have roots in several dozens or hundreds of villages, and independent beginnings on at least four continents.

> If one were asked to state briefly and succinctly what are the outstanding positive features of aboriginal civilizations I, for one, would have no hesitation in answering that there are three: the respect for the individual, irrespective of age or sex; the amazing degree of social and political integration achieved by them; and the existence there of a concept of personal security which transcends all government forms and all tribal and group interests and conflicts.[138]

A craft village is also ruled by a hierarchy of respect rather than authority. It may include members of several different families and it may have masters and apprentices, but an artisan is an artisan and, where everyone can take care of themselves, all are equal. One potter or basket-weaver may be more skilled than another, but they are all artisans.

On the face of it a master and his apprentice are not equal but an apprentice in a craft village was probably someone who pestered a master for instruction rather than someone who had to do what he was told. Even an apprentice who has to obey is training to be a master himself and is potentially equal to the master who trains him. A trader may be wealthier than an artisan, but he knows that his wealth depends on artisans.

But the officers and men of an army are not equal and, after the army takes a village, the villagers will not be equal to the conquerors. After the conquest, the commander of the army is the unquestioned ruler of the village. He will soon be transformed into a robber baron, and from now on we will refer to him as *The Baron.*

Under The Baron are his officers, stratified according to rank but all of them distinctly superior to everyone else. Under the officers are the soldiers and under them the villagers.

This is quite different from the prestige that might be granted to an exceptional individual in a hunting and gathering community or a craft village, because in a free society prestige must be earned and justified by the individual who enjoys it. In the occupied village, a common soldier of the occupying army outranks a respected villager and the officers of the army — even The Baron himself — will demand that the villager bow to the soldier.

Castes

The presence of the army will also create caste distinctions among the villagers. Traders will probably rank highest among civilians, because the army appreciates the value of trade. It may have robbed some of these same traders in the past, and it recognizes them as a source of wealth.

Traders who pay a toll to The Baron will be free — even encouraged — to come and go as they please and to bring as much trade as they

[138] Radin, Paul, *The World of Primitive Man*, Grove Press, NY, 1960, c1953, pg.11.

can to the village. In return, the army promises that it will not rob traders and that it will chase other robbers away from the area. For a price, the army may even provide traders with an escort through dangerous areas.

Some craftsmen will also have higher status than other villagers, either because they produce weapons the army can use or high-quality goods that can be taxed or traded. Either way they will be allowed or encouraged or, if need be, forced to ply their trade. In return, the army will treat them better than other villagers.

Some villagers will become tools of the army. In the first days after the conquest the soldiers may have tried to control everything directly but we can assume that did not work because it did not become the standard practice. Local villagers know their way around and invading soldiers don't, so the soldiers could not run the village efficiently.

The system that seems to work, and which is still the norm, is for the conquering army to appoint a few locals as agents to pass their orders along to the others.

Some villagers will be more willing than others to serve their conquerors but we need not assume that all the men who agree to help the army are traitors. Some might collaborate because they hope to gain power but the ideal civil administrator from the army's point of view is popular and well-respected, and such a man may accept the job because it gives him an opportunity to protect other villagers from excesses by the soldiers.

Either way, villagers who serve the army will be seen as allies of the army and enemies of many of their fellow villagers. Eventually they may begin to see the soldiers as friends and their fellow villagers as real or potential enemies.

The Lucifer Effect

We can expect this to produce the kind of behavior that psychologist Philip Zimbardo calls· "the Lucifer effect." It probably dates back to before the occupation of the village, but I suggest that it was in the occupied village that it flowered.

In the summer of 1971, Zimbardo headed a team of psychology researchers who set up an experimental 'prison' in the basement of Stanford University's psychology building, Funded by the U.S. Office of Naval Research, the experiment was intended to study the causes of conflict between military guards and prisoners.

The subjects were twenty-four male university students, all of whom were judged to be 'normal' middle-class Americans. Half were dressed and classed as 'prisoners' and the other half as 'guards' who worked in four shifts of three each.

The guards were not allowed to hit the prisoners but after a couple of days they began bullying them, insulting and demeaning them and waking them in the middle of the night and forcing them to do push-ups.

The results were shocking — not least to Zimbardo himself. The experiment was conducted in 1971 but Zimbardo found the results so disturbing, he says, that he was unable to write a popular book about it until more than thirty years later. *The Lucifer Effect, Understanding How Good People Turn Evil* describes this experiment and others, and Zimbardo's concepts of 'dispositional' and 'situational' behavior. 'Dispositional behavior' is the product of an individual's disposition, while 'situational behavior' is the product of his or her situation. The same individual can show different types of behavior at different times — as a thug or a bully in one context might be gentle with his own family.

Back in the occupied village we have to assume that the soldiers' natural disposition is neither kindly nor gentle — they are, after all, bandits and/or soldiers — but as rulers of the village they are now in the same situation as the American students who, in Zimbardo's experiment, turned into sadistic bullies.

The Baron and his officers will protect some villagers — the craftsmen who supplied arrows and weapons before the conquest and who are now seen as friends and allies, and traders who pay tolls and duties — but others are fair game and we can assume that the soldiers will rob, bully and abuse them.

We're used to this kind of behavior — it's what armies do — but at this point it was new. Bullying has been observed among chimpanzees so it's reasonable to assume there was probably some among hunters and gatherers and in the craft village, but that would have been small scale and relatively benign. A big chimp might take a choice morsel of food from a smaller one or perhaps force himself on a less-than-willing female, but there are limits to what one can do in a band and still be accepted as a member of the band.

Weaker members may not be able to throw a bully out of their band but band membership is fluid and people can move from band to band within a mating network. If they can't throw the bully out of the band, victims of a bully can move to a different band.

But the situation is different when an army takes a village, because the villagers will not be allowed to leave.

Even if some of the soldiers come from the village or from the same bands as the villagers they are now soldiers, not neighbors. They may have joined the robbers because they were rejected by the village or their band or they may have been guards of a caravan that the robbers took but, either way, they are now bandits and soldiers rather than villagers and tribesmen. They may or may not be bullies at heart and they may or may not resent the villagers who rejected them but, whatever they feel, they are now in the army and they must go along with what the army does.

In hunting and gathering bands and in craft villages, bullying would have been frowned on. In the occupied village, it is licensed.

Soldiers and Slaves

As the army settles in, some of the villagers will be enslaved. The idea of slavery is not new — some species of ants keep slaves and they may have done so for tens of millions of years before the first man was born — but we can assume that up to this point there was no long-term slavery among humans.

Hunters and gatherers don't keep slaves, partly because most of their 'work' is more pleasure than work and few people would want to avoid it.

Perhaps more important, hunters and gatherers have no way to keep or to use slaves. A hunting and gathering camp is not normally fenced, even in modern times, and everyone raised in a hunting and gathering society knows how to live off the land. Unless he or she were continuously watched or tied up, a slave in a hunting and gathering camp could simply walk away and disappear.

Further, most of the work for a man in a hunting and gathering society is hunting and to make the best use of a male slave his master would have to give him weapons and turn him loose in the wilderness. A female slave would not need weapons but, to be useful, she would have to be free to roam the wilderness and, like the hunter, she could easily run away.

Craftsmen in a village might be able to use slaves but, before the army came, it would have been difficult to keep them because the village was barely one step removed from a hunting and gathering economy and everybody would have known how to live as a hunter-gatherer. Besides, virtually everybody in the village has relatives nearby who would not allow a kinsman to be enslaved.

But an army must keep constant watch and, because it is always on guard, it can keep captives and force them to make and repair weapons and equipment, to build camps and even to fight. The idea of slave soldiers may sound strange but history and even modern times have seen many armies of slaves. The Mamluks, who ruled several countries in the Middle East for a couple of hundred years, began as a slave army, organized in AD 833 by the Caliph of Baghdad. For about 300 years, starting in the late 14th century, the Ottoman Empire maintained an army of slaves called Janissaries and, in modern times, many countries have fought wars with armies of unwilling conscripts who were, for all practical purposes, slaves.[139]

In the occupied village any villagers who are of no immediate use to the army are in essentially the same position as the losers in a game of musical chairs.

[139] Since the debacle of Vietnam, the United States armed forces have been all volunteer, but slave armies still exist. See "Child soldiers total 300,000 worldwide, UN says," *The Globe and Mail*, June 13\01, pg.A12.

Once a popular school and parlor game, 'musical chairs' is played by a group of people who dance around a circle of chairs. When the music stops they all sit down but there is always one more player than there are chairs. One person is left standing each time and he or she drops out of the game.

Villagers who do not have an established role when the soldiers come don't have the option of dropping out. They must accept the role the soldiers assign them, and that role may be as slaves.

The first job for the slaves will be to build a fortress. It will be a huge building by village standards because many of the soldiers will have to live in it, perhaps for the rest of their lives, and it must be able to hold out against a long-term siege.

This is a new requirement because bandits spend most of their time either running or hiding or pretending that they are not bandits. After the army takes the village everyone knows who and what and where the soldiers are, and most of the villagers are their enemies. Hunters and gatherers who live outside the village are friends and relatives of the villagers and therefore enemies of the soldiers.

To be safe the soldiers must live in a fortress with a store of food and a well. The village may not have a well if it lies beside or perhaps straddles a river or stream, but even some animals know that if you dig a hole you often find water. We have to assume that men probably knew about wells by this time and, in case of siege, the soldiers will demand a well within the fortress.

We don't have to assume that all conquering armies built fortresses, dug wells and stored food against a siege, but it is a safe bet that most of the ones that survived did.

The fortress will be the biggest and strongest structure in the village and, through history, fortresses and castles have been among the most impressive structures produced by any culture. Some fortresses and castles that were built more than a thousand years ago are still in use.

At this point most buildings are probably built of sticks and wattle, or perhaps mud brick. The fortress might be built the same way but the idea of stone construction is probably known, if not well developed and, with captive villagers to do the work, the soldiers may demand that their fortress be built of stone.

The army may bring slave girls to the village, or enslave some village women, and this may be the beginning of the 'battle of the sexes' that causes so much misery in the modern world.

Among hunters and gatherers men and women are inherently equal, because either could live without the other. Among most groups there is a division of labor — men hunt and women gather — but that is a matter of convenience and custom.

A single man could do his own gathering and a single woman could trap or snare small game. In fact she would not have to because, among most hunters and gatherers, everyone in a camp is entitled to a share of a big kill, whether they took part in the hunt or not.[140] The only time women in a hunting and gathering society really need men is when they are in the last stages of pregnancy or have young babies to take care of, and if they have no men they don't get pregnant.

Men and women in a craft village are equal too. They may work at different crafts but one is as important as another and both men and women could go back to hunting and gathering if they had to.

But the women who follow an army are not free. Some camp followers are probably volunteers but some are probably slaves, because raiders who capture a camp or a village can take any women who survive the raid. As captives these women will have no rights and they can be treated as property. Because slave girls have no rights, all women who follow the army have to do what they are told.

The village women most likely to be taken as slaves will be the widows and daughters of men who died when the raiders came, but women whose husbands or fathers survive will also lose their freedom. They are now hostages and, even if they are allowed to live at home, they are liable to be killed or enslaved if their men rebel against the new rulers.

We see an echo of this in some Muslim countries where women are not allowed to appear in public without a male to protect them. In the modern world this is a formality rather than reality, and in some areas, a five- or ten-year-old boy is seen as adequate protection for an adult woman.

In a village ruled by soldiers, women are property. They are subject primarily to the whims of soldiers, but male domination will be the norm.

Soldiers will dominate women because most of the women they deal with are slaves and, after the army takes over, the men of the village will dominate women because the soldiers do it. They must, because if they do not they give the soldiers another point of superiority over them.

They can do it because the threat of the soldiers gives the village men power over the women of their families. Among hunters and gatherers or in a craft village women can leave if they don't like the way a man treats them but when soldiers take over a woman is trapped. A wife who leaves her husband or a daughter who leaves her father's house has no protection and may be taken as a slave by the soldiers.

Women who are not taken as slaves of the soldiers may be forced to collect food. When it attacks the village, the army probably has food for a couple of days but not much more, and it is now surrounded by enemies — inside the village and out.

Even with the army to feed the combination of hunting, gathering and trade might have been enough for the occupied village but when it

[140] Even now, in the Inuit village of Rankin Inlet on the shore of Hudson's Bay, a successful hunter often reports his kill to the local radio station and broadcasts an invitation for anyone who wants a share to come and get it.

occupied the village the army made enemies of the surrounding bands — whose friends and relatives live in the village — and there will be less trade. More, when the army took the village it made enemies of the surrounding bands, who were relatives of many of the villagers, and for the bands, it might make sense to burn the fields around the village in hopes of starving the army out. Whether the bands planned to burn the fields or not, it made sense for the soldiers to guard them and to keep watch over the village women who were sent to gather. As they guarded the fields, the soldiers would have appreciated their value, and the officers might have claimed them as private property and forced the villagers to develop them as farms.

The Cash Economy

The conquest of the village also created the cash economy. Until that time most villagers, like hunters and gatherers, lived in what economists call a 'subsistence economy.' Some think that means they were poor, but that's not necessarily so. A subsistence economy is one in which people make, collect or hunt most of the things they need, and they do not have to trade to live. They may be rich or poor but they do not live by trade and they don't try to amass wealth that they have no immediate use for.

In the Middle Ages a noble whose estate included farms and had its own construction crew, smithy and other services seldom had much money and, technically, lived in a subsistence economy.

Even now most hunters and gatherers in subsistence economies have a higher standard of living than most of the farmers and industrial workers of history. They do not have television or washing machines but they don't need them and they work less, play more and eat better than most of the modern American middle class.

At the other extreme, consider a child laborer knotting rugs or sewing soccer balls in a third-world village. He or she may work 15 hours a day and earn just barely enough money to buy the cheapest food available but, because he or she must earn money to buy food, he or she is part of a cash rather than a subsistence economy.

Even in a subsistence economy there was probably some trade, but it was not the basis of the economy and it would have been conducted in a very different manner. Except for luxury or exotic goods like jewelry or axes, most trades would have been with friends and relatives and the traders would probably have granted each other almost unlimited credit.

If I buy a pot from your wife, I might pay for it by giving a fish to your brother or arrows to your cousin, sometime in the next year or so. We are all friends and relatives and, in an economy of plenty, we don't have to worry about precise book-keeping.

Even among bands and villages that supported full-time artisans the most dependable store of value was in family ties, friendship and general good will.

But there is no good will between the army and the village it has conquered and, even among villagers, friendship and family ties are no longer as dependable as they once were.

We can assume that there was no money in the world at this point but, even so, this is the beginning of the cash economy. Formal money is a relatively recent development but, by this time, the concept of treasure would have been well established.

We know that jewelry dates back to the Neolithic because some has been found. We can infer the possibility that Neolithic axes may have been used to store value because, as noted earlier, some of the earliest ingots of copper in northern Europe were cast in the form of a double-bitted axe — perhaps to suggest a relationship between the value of the copper and the value of an axe. In the Mediterranean area, copper ingots were often cast in a shape reminiscent of an ox-hide. In historic times tea leaves, cocoa beans, sea shells and huge rocks have all been used as forms of money.

We can't guess what the people of the occupied village used as a medium of exchange but we can assume that people who were free to demand payment for their goods or services would do so. Whatever the currency, the occupied village will have a cash economy.

This is a two-edged sword. On the one hand a cash economy allows the division of labor to an incredible extent.

In the opening pages of his classic *Wealth of Nations*[141] Adam Smith describes a workshop in which ten men make pins. One man draws the wire, another cuts it, another puts a point on the pin, another adds the head, yet another packages the finished pins and so forth. Between them, Smith says, they make about 48,000 pins a day — far more than they could if each man did every job.

Smith's argument is valid, but if he wanted to count all the men who had a hand in making those pins he should have included the men who made the steel, the men who mined the ore and coal that was used to make the steel, the men who made the tools that were used to mine the coal and the ore, the men who made the axe that cut the tree that was used to make the paper the pins are packaged in, and so forth. In a cash economy work can be divided to the point where literally thousands of people contribute to the production of a simple pin.

Even at the level of first-remove it takes literally dozens of factories to make a complex product. According to one industry estimate, for example, the average American car is made in about 100 separate factories — and that is very conservative.

141 Adam Smith, *Wealth of Nations*, first published in 1776, republished by Encyclopedia Britannica, Chicago, 1952.

The engine is made in one plant, the body in another, the tires in another and so forth, and each one of these components is itself the product of several factories. The finished engine, for example, will include castings from a foundry, machined parts from several machine shops and filters, wiring, an electrical starter motor and alternator, spark plugs, fan belts and dozens of other parts that come not only from different factories, but also from different companies and often from different continents. If we split it down further we would find that the castings are made from steel that is produced by a steel company from iron that was mined by a mining company with machines made by a machine producer and transported by ships made by yet another company, and so-on.

A starter motor includes wiring that is made by one company from copper produced by another and insulation produced by yet another.

The cash economy makes it possible for countless people who will never meet each other to cooperate in the production of goods that could never be produced in a subsistence economy, but it has drawbacks. One is that cash is an abstract form of wealth and, in a cash economy, we can and may be tempted to hoard beyond reason.

Hunters and gatherers can store food but, in an economy of plenty in which food is cheap and storage is relatively expensive, they do not store much. Once they have all they can use, they quit working.

But we can never have too much cash. Wealthy men around the world who have far more than they can ever consume keep working. In some cases they love their work but in others the love of money, or the perceived need for an infinite supply of it, is a form of madness. Some of them increase the general wealth of the community by their work but many of the wealthiest people in our society take wealth from others and produce nothing themselves.

The concept of hoarding probably began about this time. An army must be prepared for a siege and it needs to store enough food to supply the soldiers and some of the villagers for an indefinite period. Besides, where food is bought and sold a store of food is wealth.

The combination of stratification and the potential for hoarding also create the potential for conspicuous consumption. Because a person's comfort and even his life in the occupied village may depend on his or her position in the hierarchy, people feel a need to attain and display high status.

Hunters and gatherers enjoy high status, too, when they can earn it, but where wealth can't be easily stored a man displays his ability to hunt, gather or make more than he can use by giving the surplus away as gifts. In a cash economy wealth can be stored or invested in treasure that can be displayed, and a display of wealth is so powerful that criminals and others who are known to be enemies of society may be 'respected' for the opulence of their public display.

In the craft village there was no need for display and, even though they may not have looked prosperous, the villagers must have had an easy life. We can presume this because we know that hunters and gatherers had an easy life and, in a free and un-crowded world, people could live where and how they wanted to. If some chose to live in a village we can assume that life in a village was as good as life as a hunter-gatherer.

But that was partly because everyone was self-supporting and no one had to provide for more than himself or herself and his or her family. In the occupied village the villagers have to support the army — which may out-number the villagers — and the villagers who collaborate with the soldiers and the slaves who build the fort and, beyond all this, they must also provide the army with a surplus of food to store against a siege.

Further, the people of the craft village were free to hunt and gather in the fields around the village, and neighboring hunters and gatherers were free to come into the village to trade. After the army takes over the villagers are not allowed to gather, except under guard, and hunters and gatherers do not come to trade. Now the army controls the food supply and even the women who gather food can eat only what the soldiers allow them to keep. Soldiers can hunt near the village, but they don't have to share their kill with villagers.

History tells us that many early peasants and villagers lived in poverty and that even petty nobles lived in relative magnificence. Our culture teaches us to regard the nobles' wealth as the property of the nobles — but not to ask where it came from or how they got it.

Makers and Takers

In *The Forest and the Sea*, biologist Marston Bates suggests that life forms can be sorted into three groups that he describes as *producers, consumers* and *decomposers*.

Producers, mostly plants, produce food — mostly by photosynthesis. *Consumers*, mostly animals, eat either producers or the fruit or seeds they produce or, if they are carnivores, other consumers. *Decomposers*, mostly bacteria, decompose dead *producers* and *consumers*.[142]

I suggest that human economic functions can be sorted into two distinct groups, each with two sub-groups. One person might conceivably be a member of all four categories at the same time, but still, the groups have different characteristics. I call the two main groups the *makers* and the *takers*.

Makers make, or hunt or gather or grow, the goods we use. Among hunters and gatherers everybody was a maker and, as a general rule, each family or small band made or hunted or gathered everything it used. Among some groups, full-time artisans made goods for others and accepted the fruits of others' hunting and gathering — and perhaps goods made by other specialized artisans in trade, while Makers in the modern world work in craft shops and in factories, and

[142] Bates, Marston, *The Forest and the Sea*, New American Library, NY, 1961, pg.117.

on farms, ranches and fishing boats. They are the people who support the rest of us.

Takers take things from the makers who produce them. Robbers are takers and so are The Baron and his army. Because they rule the village, we do not call the soldiers and their collaborators robbers, but they produce nothing themselves and they take what they want by force or by threat of force. Even the commander and soldiers of a defensive army are takers, because they can demand what they want and the villagers must provide it. If they don't, the army could take it anyway — or it could quit and leave the village open to take-over by another army.

We also have two sub-sets. *Traders* live on the produce of others but they don't take goods without payment and they perform a valuable function because goods gain value when they are taken from one place to another — even if only from the workshop to the market. Even though they don't make things, we have to see most of them as a sub-set of makers.[143]

In Neolithic times, top-quality stone, blanks and finished axes were traded over hundreds of miles. Even if there were no robbers, the traders' life was sometimes dangerous but they brought axes to people who had no access to the quarries that produced the best stone, and they spread ideas.

Agents include villagers who might help makers locate raw materials or who mind the traders' business while they are out of town, but these are a small group. They can be lumped in with the makers and traders they serve. More important, in the long run, are the villagers and others who work for The Baron and other takers as tax collectors and so forth. I see them as a sub-set of takers because, while they may not take goods in their own name, they do take them.

This group also includes any shamans and priests who are supported by an army or who demand a 'tithe' or other tax from the villagers. At some times and places, priests were the ultimate — and often the greediest — power in the land. Agents perform a service but, when they serve takers, they must be considered a sub-set of takers.

The advent of takers and their agents brings a major change to the village. Up to this point, makers had an easy life. After takers move in, the makers have to support them and their agents and, with The Baron and his soldiers as a model, some villagers will try to become takers themselves by collaborating with the invaders or in other ways.

[143] As we make these distinctions, we must remember that they describe functions, not people. A soldier is a taker while he is on duty, but when he is off-duty he might make arrows or hunt or gather food, which he might sell in the market. If he also works occasionally as a tax collector or administrator, he might at different times be a member of all four groups — a taker, a maker, a trader and an agent. If we were trying to judge people this man would be difficult to categorize but because we are trying to describe economic functions we have no problem. At any given time, he is what he is at that time.

Before the army took the village takers were, by definition, criminals. After the conquest takers made the rules and they made some forms of taking legal. For the past several thousand years many of the brightest and potentially most productive people of every era have spent most of their lives developing new and marginally legal ways to take the products of makers.

Our modern economy is so complex that the line between makers and takers is blurred, and it's blurred still more by the fact that a modern economy could not function without some people that we might consider to be takers; but we're not talking about the modern economy. This was a village that was taken by an army before the dawn of recorded history and, at that time, virtually all takers were parasites who lived off makers.

A Community of Friends and Enemies

The occupied village must have seemed strange to people of the time, because up to this point all human communities had been groups of friends, relatives and allies. The members of hunting and gathering bands, the residents of villages and even groups of bandits all have common aims and ambitions, and all are willing to protect and help any other member of their group. We have already quoted anthropologist Paul Radin on the aboriginals' "concept of personal security which transcends all government forms and all tribal and group interests and conflicts."

But after the army moves in, the village is more like a coral reef where many different kinds of fish and other animals live together; and some are food for others. Neighbors in the occupied village don't eat each other but they are not all friends, and some may exploit, rob or kill others.

The Baron rules but he knows that many of the villagers and even some of his own men would like to kill him, for revenge or in a bid to take over. The officers of the army live well but every one of them knows that if he drops his guard he might be murdered by a villager, by one of his own soldiers or even by another officer who sees him as a rival. They are in the same army now, but some of them were once deadly foes.

During the war in Vietnam there were rumors that the United States lost nearly as many officers to their own men as to the enemy. In the opening days of the Second Gulf War, two American officers were killed and 14 wounded when an American soldier rolled a grenade into the tent in which they slept.[144] On November 5, 2009, U.S. Army major and psychiatrist Nadal Hasan fatally shot 13 people

144 "American soldier detained in grenade attack on US troops," *HoustonChronicle.com* Mar 22/03

and injured more than 30 others at Fort Hood, a U.S. Army base near Killeen, Texas.[145]

Soldiers in the occupied village don't have to fear the villagers when they stay in groups, but any one of them might be in danger if he were to walk down a dark alley alone or go alone into the woods. A soldier also has to watch out for officers who might catch him in some breach of the rules or who might assign him extra duty or some other inconvenience.

Traders know that soldiers might protect them from robbers on the road but, if the opportunity arose, those same soldiers might rob them. With the advent of the cash economy and the pressure of conspicuous consumption in the village, traders also have to worry about villagers who might pilfer their wares or steal their treasure.

Some of the craftsmen in the occupied village will be wealthier than they were in the craft village, but they won't have as pleasant a life. Like hunters and gatherers, free craftsmen work when they feel like working but, in the occupied village, a craftsman must ply his trade full time or risk being enslaved. Further, the residents of a free village can hunt and gather their own food, but now The Baron and his officers control the land around the village and they do not allow villagers to hunt or gather for themselves. Craftsmen who make equipment for the army or expensive trade goods may do very well — The Baron may even give them slaves to help with the work — but they must fear the competition of other craftsmen, the officers and administrators who may demand bribes, their own slaves and other villagers who may rob them.

Villagers who serve as agents of The Baron live in physical comfort but also in fear— of The Baron and his demands on the one hand, and of possible reprisals by villagers on the other. Slaves may or may not live in reasonable comfort but, either way, they will resent the soldiers who enslaved them and the masters who use them.

Like hunters and gatherers, the residents of a craft village lived among friends. Like the denizens of a coral reef or of a modern city, the residents of an occupied village lived among enemies and potential enemies.

The New Order

We also have the beginning of poverty. This is a new development because up to this point, living was easy. People worked as much or as little as they chose to, at tasks they chose themselves, and whether they worked or not everyone had plenty to eat.

Under the army most of the villagers will be poor and everybody will have to work to eat. Even the conquerors and their collaborators have to work harder than free hunters, gatherers or craftsmen.

---

[145] https://en.wikipedia.org/wiki/2009_Fort_Hood_shooting

The Baron gets first choice of food, but to protect his position and perhaps his life, he has to please, pay and dominate his army and his officers. The officers in turn must dominate their men and please the Baron and, soon, a group of army officers, village collaborators, some traders and others will form a 'court' in which they will plot to gain the Baron's favor and watch out for others plotting against them. Executives of many modern corporations would feel right at home.

The army and its collaborators live on the work of the villagers, and we could consider them parasites, but that does not mean their life is easy. Even low-level minions in the Baron's service must make a display to show that they have the Baron's favor and, because they will be judged on it, they are under continual pressure to make more and more magnificent display. Most of the minions will be wealthier than most villagers, but they will need all their wealth and more to pay for the display they need to maintain their position. For his part the Baron will allow his favorites to take some villagers as slaves or to demand taxes, tolls or some other type of payment from them.

If the Baron expands his domain he may give some of his officers villages to rule in his name and, if he gives some villages to his supporters, he will have to conquer others to maintain his own wealth. In time, he will become a king or an emperor and, for most of the period we think of as 'history,' most of humanity has been enslaved by a few rulers and nobles. It is tempting to think of the rulers and nobles as villains — and I do not suggest that they are blameless — but we must remember that they could not rule alone, and they were not much more free than their subjects.

For every king, there were dozens of nobles who supported him and who expected him to act like a king. Every noble was supported by officers who expected nobles to act like nobles and the officers were supported by soldiers who expected officers to act like officers. The system also included tax gatherers who were expected to act like tax gatherers, priests who were expected to act like priests and so forth.

Slaves and serfs are also part of the system, because they know no other way to live. Even now, in what we consider an age of enlightenment, experiments show that most well-educated middle-class Americans will obey commands that we might expect intelligent and humane people to resist.[146] In the village, freedom will soon be forgotten and The Baron and his friends — and even the villagers — will accept The Baron's right to rule.

Robbers and armies take what they want with threats and violence but, under The Baron's rule, new forms of taking developed.

The Baron himself started the change because, like most conquerors since, he found some villagers willing to collect the tribute that we now call taxes for him. In historic times collaborators kept records of taxes collected and listed craftsmen and their skills and so forth.

[146] Milgram, Stanley, *Obedience to Authority*, Harper & Row, 1974.

Because many of these people are resented by the villagers, they form a metasystem among themselves, and with the army and the developing MIC.

There was no need for record-keeping in the craft village because every individual and family was self-supporting and, in a village of neighbors, there was no need for civic works. Villagers built their own huts, perhaps with help from friends and neighbors. They made their own fire-pits, and they might agree on and perhaps build a designated latrine. The village probably lay beside or even spanned a creek or a small river and people who crossed it often might have added stepping stones to make the crossing easier. They would all have drawn water from a convenient spot on the river or creek bank and, gradually, the spot might have been tailored to make it easier to draw water. Like campsites on a canoe route or hiking trail, the village would be gradually improved by users.

But The Baron, his household and his army are not self-supporting and the villagers are now required to provide housing and food for the army, a castle for The Baron and a palisade or wall to protect the army from its enemies and make it difficult for villagers to leave without the army's permission. The villagers didn't need a well, but the army would demand one, inside the wall, in case of siege.

In time, the occupation of the village will produce a caste system because where you have one class of people living as parasites on the produce of others, you need a third class to make sure the workers work and that the top level of parasites get the benefits of that work. There were probably some agents in the village before the conquest, because a trader might have an agent in his home village to manage his business while he is away and agents in other villages to find products for him to buy and customers for the goods he will bring. The bandits may have had agents too — to buy supplies for them if they are not welcome in the village and to tip them off when a rich caravan is about to leave.

After the conquest The Baron and the army will use these and other agents to supervise the villagers and perform other administrative functions. Agents are not actually takers but, if they represent takers, they must be seen as takers themselves.

The New Takers

In the modern world we don't view civil servants, lawyers, accountants and other people who do not produce 'real' goods as takers but, even though many of them perform useful functions, it is obvious that they would all starve in a country without makers and that a country of makers without these people could do well.

In an earlier book[147] I distinguished between what I call *benefit* and *cost goods*. Both categories include both goods and services, and

[147] Turnbull, Andy, *The Numbers Game*, Red Ear, Toronto, 2001.

the distinction between them is that if production of *benefit goods* increases, the overall cost of living will go down but if the production of *cost goods* increases, the cost of living will increase. In general, farm crops and other real products are *benefit goods* and most government services are *cost goods*.

While I make this distinction I do not pretend that our modern world could work without governments and bureaucracies. Still, the fact remains that most bureaucracies produce cost goods and it seems to be their nature to grow. Historian C. Northcote Parkinson argued this in his best-selling *Parkinson's Law*.[148]

In an amusing analysis which has been cited countless times since, Parkinson reports that in 1914 the British Royal Navy had 542 ships, 125,000 officers and men, 57,000 dockyard workers, 3,249 dockyard officials and clerical staff and 4,336 Admiralty officials and clerical staff. The number of ships, officers and sailors decreased over the years but the Admiralty staff increased and By 1967 the navy had only 114 ships, 83,900 officers and men and 37,798 dockyard workers, but 8,013 dockyard officials and clerical staff and 33,574 Admiralty officials and clerical staff.[149]

Parkinson argued that one reason for the increase is that officials want subordinates, not rivals. If a pompocrat has only one assistant, the assistant will have to know as much as the pompocrat and could replace him. Instead, the pompocrat will insist on having two assistants, each of whom will know only half the job and neither of whom could replace him.[150] The problem is that if the bureaucracy lasts long enough, the officials will retire, and they will often be replaced by assistants or deputies that were chosen for lack of intelligence, vigor or a pleasing personality.

Even the first generation of officials can't afford to be too efficient. More than fifty years ago, when I worked as a newspaper reporter, I discovered the principle that I call *Turnbull's law*. When faced with a problem and a selection of possible solutions, the ideal choice for a civil servant is the most expensive program that will not solve the problem.

The logic of this is obvious. The more expensive the program, the more important the person who chooses and implements it — but if the program actually works, the problem disappears. The civil servant who battles an insoluble problem may have a long and prosperous career, but if he or she were to find a cheap and efficient solution to that same problem, it would be assumed that the problem was not serious and the civil servant might fade into obscurity.

In 2010, the U.S. Census counted a total population of 308,745,538 people, of whom 2.79 million worked for the federal government

148 Parkinson, C Northcote, *Parkinson's Law* Houghton Mifflin, Boston, 1957. My notes are from the 28th printing.
149 Ibid., pg.4.
150 Ibid., pg.17.

and an estimated 24 million worked for state, county and city governments.[151] I don't suggest that the United States could operate without government, but 2.7 million people plus another 1.2 million lawyers, 1.2 million accountants and auditors and 1.58 million book-keepers adds up to a lot of non-producers. Now, consider that most of these people are paid more than the average maker earns, and you get an idea of the kind of burden The Baron, and governments through history, have imposed on the people who support them.

In New York State, for example, the average income in 2004 was $40,272.29 per year[152] while, according to the Vera Institute of Justice the average cost of a prisoner in 2010 was about $60,000 a year[153] and a story in *The New York Times* reports that in 2013 prisoners in New York City cost about $168,000 per year.[154] I suspect that most prisoners would be willing to retire to an honest life in Florida if the money were paid to them directly.

We need civil servants in the modern world but it is more important, from a civil servant's point of view, to understand the workings of office politics in their departments than to understand the jobs they are supposed to do or the industry they're hired to regulate.

While I was working as a news reporter in central British Columbia, the city hired an 'industrial commissioner' to bring new industry to town. This was at the beginning of the electronics boom and this city — remote from markets but a wonderful place to live with crystal-clear mountain air and a post-secondary college — would have been an obvious place to locate factories producing electronic goods.

But when I interviewed the newly-hired 'industrial commissioner,' he had no ideas about the type of industry he should try to attract. In an attempt to draw him out I made some suggestions and, in the end, he agreed that the city would be a wonderful place to build a foundry to cast anchors for the navy.

We had no iron in the area and no coal suitable for use in a foundry; we had no trained foundry workers and any anchors cast in that city would have to travel hundreds of miles through mountains to reach salt water. On the other hand, a foundry would have polluted the clean air that was a valuable asset.

[151] https://www.census.gov/2010census/data/

[152]http://www.google.ca/search?qaverage+income+ny+state&hlen-CA&gbv2&oqaverage+income%2C+ny&gs_lheirloom-serp.1.2.015j0i3015.81439.89777.0.92233.18.16.0.2.2.0.147.1563.11j5.16.0....0...1ac.1.34.heirloom-serp..0.18.1570.elawELcSakM

[153] http://www.google.ca/search?qcost+of+average+prisoner+per+year&hlen-CA&gbv2&oqaverage+income+ny+state&gs_lheirloom-serp.12...0.0.1.401409.0.0.0.0.0.0.0.0..0.0....0...1ac..34.heirloom-serp..0.18.1570.LHE6tLjaRt4

[154]http://www.nytimes.com/2013/08/24/nyregion/citys-annual-cost-per-inmate-is-nearly-168000-study-says.html

The industrial commissioner didn't have a clue about the job he was hired to do, but he did know how to work the system and, in time, he became one of the most powerful bureaucrats in the city.

When housing prices in that city began to soar, I wrote a series of articles on the problem for a local newspaper.[155] Developer Frank Hewlett told me the increase was largely because of city planning.

To illustrate his point he phoned to check the price of a ten-acre lot that was advertised for sale. The owner wanted $324,000, and if Hewlett would put $100,000 down, the owner would carry the balance at 10 percent. That was a reasonable price for the land, at that time and place, and 10 percent was a reasonable rate of interest for a solid developer with unquestioned credit. Hewlett figured he could put 44 houses on the land so, without interest, the lots would cost about $7,300 each. That sounds cheap now, but this was at a time and place where you could buy a nice house for $15,000.

And the cost of the land was just the start. Hewlett figured it would take about 13 months to get the paperwork sorted out, the services engineered and the project approved by city hall. The paperwork would cost about $8,800 per lot — more than the land itself — and when it was finished Hewlett would have to pay the city another $2,000 per lot for off-site services.

The total cost for the land, the paperwork and the city connections would be $18,100 per lot in a city where, at that time, finished houses were still selling for $15,000. Because of a new planning bylaw, the cash value of every house in town increased by at least half, and many doubled or tripled.

About this time a friend of mine sold his house for $15,000. In less than a month it was re-sold for nearly $25,000.

As a writer for and editor of trucking magazines, I have been appalled by the ignorance of the civil servants who write and enforce trucking regulations and agreements.

More than thirty years ago I wrote several articles about Transport Canada's plans to mandate front-wheel brakes on big trucks. The bureaucrat/engineer who promoted the regulation had a demonstration set up, with a small ramp and a wooden model of a tractor-trailer truck. He showed me that the combination could jack-knife with the tractor's rear wheels locked but not with the front wheels locked and the rear wheels free to turn.

He said his demonstration proved that the truckers' worry that front wheel brakes would cause rigs to jack-knife was unfounded. That was a surprise to me because by that time I had spent about five years talking to truckers and writing about trucks, and no one had ever told me that front-wheel brakes would make a rig jack-knife.

[155] Turnbull, Andy, "Spotlight on housing #2, City land prices a major problem," *Kamloops News*, Dec 8/75, pg. 22-23.

When I researched this story, I spoke to working truck drivers and while some had heard that front-wheel brakes might cause a jack-knife, none of them worried about it.

Most Canadian trucks already had front-wheel brakes, but there were a few truckers who did not want them. They drove in hilly country on roads that were sometimes slippery, and if they did not make it to the top of a hill, they would have to back down — but if they had front wheel brakes, very little brake pressure would make their front wheels lock up and they would lose all steering control. On the other hand, if the hill was so long and so steep that they could not make it up on the first try, it would be deadly dangerous to try to back down without using their brakes.

Front wheel brakes also make it impossible to brake and steer at the same time on a slippery surface. Logging truck operators worry about that because they often drive on slippery, muddy roads with steep hills and sharp curves. They have to use brakes to slow down for curves, but if their front wheels lock up, the truck will go straight ahead — and on some logging roads that could drop you over a hundred-foot cliff.

Most truckers could accept front-wheel brakes with a cutout switch to shut them off when necessary, but the proposed regulation would not allow that. Because the brakes were mandated, truckers who disconnected them invalidated their insurance.

In the USA, at the time, all new trucks came with front-wheel brakes, but truckers were free to install a cutout switch or to disconnect them completely.

I wrote these reservations into my story for Transport Canada's *Transpo* magazine but — presumably because it showed that a senior bureaucrat didn't understand the implications of the regulation he was imposing — the editor took them out and the regulations were approved. Since then I have seen several trucks that went off the road because they could not back down a slippery hill with front-wheel brakes and, in fact, I was riding in one of them at the time. Because this is more likely to happen in remote areas (I've seen it on roads and ice roads in the Yukon and in Labrador), running a truck into a ditch may cost tens of thousands of dollars and risk the driver's life.[156]

Last I heard, officials of the Canadian Department of Transport still did not understand why Canadian truckers are not pleased with the agreement that regulated the operations of Canadian trucks in the United States and American trucks in Canada.

One reason is that while the United States is a developed country, Canada is essentially a strip mall. Quite a few cities in the States — like Miami, New Orleans, Dallas, Los Angeles and others — are a long

[156] Because the editor changed my story I took my name off it and, if I kept a copy of it, I can't find it now. I included the truckers' reservations in "Front Brakes, a Wise Choice?" in the June/86 issue of *Trucking Canada* magazine.

way from the Canadian border, while few Canadian cities are more than a few hours' drive from the American border.

The agreement allows truckers from either country to haul from one country to the other and to make one "repositioning" move within either country if that move puts the truck into position to pick up a cross-border load. If a Canadian trucker hauled a load from a Canadian city to Miami, for example, he could then take a load from Miami to Atlanta if he had a load arranged from Atlanta back into Canada. An American trucker who hauls a load to Toronto, say, could then take a load from Toronto to another Canadian city where he would pick up a load to the States.

That means that any American trucker who hauls into Canada has no worries. Because Canada is essentially a dependency of the United States, there is never any shortage of return loads, and even if there were, most Canadian cities are only a few hours' drive from the border and from there an American trucker can take any load.

He might find it easy to get a load from Canada to the States, in fact, because Canadian truckers have to think twice about hauling to an American city more than a thousand miles from the border. Even if he can find a return load, the American shipper knows that the Canadian trucker will have to take it or run empty for a thousand miles or more, and he need not offer a fair price for the trip.

Enforcement is also a problem for Canadian truckers. The manager of the Toronto terminal of a big American company once told me that his company 'repositions' about thirty trucks a day between Toronto and Vancouver. They had no authority to haul between Toronto and Vancouver but they had trucks and terminals in Toronto and Vancouver and loads to the United States from both Toronto and Vancouver — so they sent loaded trucks from Toronto to Vancouver to pick up loads for the States, and from Vancouver to Toronto to pick up loads for the States. It's legal under the regulations but, in fact, the American company was running a large, unlicensed, trucking operation within Canada.

On the other hand I know of one Canadian company that had a rush load from eastern New York state to Winnipeg. The truck broke down near Chicago and, because the load was rush, the company sent another truck from Canada to pick up the trailer. The driver of the first truck was charged, and fined, for hauling a load within the States.

You might wonder how Canadian officials could have accepted such an agreement but I found out about that at a meeting of trucking regulators in Quebec, in 1989. I was not wearing a press badge but I was in the front row of the audience and a woman beside me assumed that I was a civil servant. This was in the heyday of the women's movement and this woman informed me with great pride that she had negotiated the international trucking agreement.

It seems her American counterpart was also a woman, and because it was two women negotiating, the Canadian woman told me she had been able to accept every suggestion the American woman made. She was proud that the agreement had been negotiated without argument, but in fact she didn't seem to know much about Canada and she didn't know or care much about trucking. She was there to score points for women, not for Canadian truckers.

At a different public meeting a former official of Fisheries and Oceans Canada told the audience why Newfoundland fishermen are less than enthusiastic about the Canadian government and Nova Scotia fishermen. He explained that most Nova Scotia fishermen have bigger boats than Newfoundland fishermen and they catch more fish, and the Newfoundland fishermen are jealous.

He didn't seem to know about the Bond-Blaine Treaty of 1890, which would have allowed Newfoundland fishermen to land their catch duty-free in the United States and American fishermen to buy bait and have their ships repaired in Newfoundland. At the time Newfoundland was one of the richest areas of North America, with thriving fishing and ship-building industries. The treaty would have made it even richer, but at Canada's insistence, Britain refused to ratify the treaty and, more, banned Newfoundland shipyards from repairing and Newfoundland merchants from selling bait or other supplies to foreign ships. This ban was a major factor in the bankruptcy that forced Newfoundland to join Canada in 1949.

I lived in St. John's for nine months in the late 1950s and every time a storm hit the Grand Banks — several times while I was there — fishing ships from the United States and several European countries sheltered in St. John's Harbor. Several times I saw it so full of ships that I could almost walk across it, from ship to ship. Every ship always needs repairs and supplies and most of these had been away from their home port for months, but while they could shelter in St. John's Harbor, sometimes for several days, local merchants and shipyards were not allowed to sell anything to them.

Of course you can't expect an official of Fisheries and Oceans Canada to know much about either fisheries or oceans. It's more important for him to know which of the senior officials in his department he has to butter up and which juniors have to butter him up.

As a Canadian, I know more about the stupidity of Canadian civil servants than American, but one 'expert's' proposal, made as part of the 'War on Terror,' is worth mentioning even though it was never implemented.

Many people think the air brakes on big trucks are applied by air pressure but 'spring brakes' are applied when the pressure is reduced. That's a safety feature because if a truck's air line is broken, the brakes will 'dynamite' — come full on.

One 'expert' who was worried about the possibility of a truck full of gasoline or explosives ramming into a government building came up with the idea that the brake lines on big highway trailers should run through a shear which would be connected to the trailer's back bumper. Police could then stop a runaway or suspect truck by bumping the back of the trailer and setting the brakes.

Great idea, except that anybody can bump the back of a trailer, by intention or accident and — accidents aside — this would make it easy for terrorists and/or pranksters to cause havoc by bumping the back of fuel tankers and other trailers carrying dangerous goods at the right time.

Just imagine the delays if a few big trucks get bumped — and are stopped with brakes that can't be released until the severed air line is repaired — on a busy road in rush hour, or on a slippery curve where a trailer with locked brakes would slide off the road. Fortunately, saner heads prevailed and the idea was never implemented.

Lawyers

In the modern world, we have a special class of agents, called lawyers, who interpret and argue over the law. Some work for government and some for big corporations, but many are 'freelance' — working for individual clients who may hire them to argue a specific case.

The United States is generally considered to be the most litigious country in the world, and while that is good for lawyers, it is a burden for other citizens. *Boston Globe* columnist Jeff Jacoby wrote on May 9, 2014: "Chief Justice Warren Burger predicted 35 years ago that America was turning into 'a society overrun by hordes of lawyers, hungry as locusts.' At the time, the population of attorneys in the United States had surpassed 450,000, and law schools were graduating 34,000 new ones each year. By 2011, the annual production of law degrees was up to 44,000, and at 1.22 million, the number of lawyers in the country had nearly tripled. Over the same period, the population of the United States had risen just 40 percent."[157]

Some lawyers must perform useful functions, but some are just plain destructive. In January of 1992, the U.S. Food and Drug Administration imposed a moratorium on use of the silicone breast implants produced by the Dow Corning company. The company wasn't worried because the implants represented only about one percent of their business and they carried $250 million insurance, but they under-estimated the viciousness of the lawyers who swarmed like piranhas on the bleeding carcass of a dead cow.

Studies by the Mayo Clinic and other reliable medical facilities found that women with breast implants have about the same incidence of diseases as women without implants but, with 750,000 implants in use, a lot of women with implants also had diseases.

Lawyers encouraged any woman with an implant and a disease to blame the disease on the implant. Lawyers across the country

[157] https://www.bostonglobe.com/opinion/2014/05/09/the-lawyer-bubble-pops-not-moment-too-soon/qAYzQ823qpfi4GQl2OiPZM/story.html

advertised in local newspapers to find women who had implants and who might be persuaded to file claims against Dow Corning and hundreds of thousands of women joined the rush, hoping for a share of the pie.

Because truth is no defense against lawyers, Dow Corning offered a 'global' settlement of $105,000 to $1.4 million to each claimant, depending on her health and age. That would have cost the company more than $4 billion but it would have limited the lawyers' take, and many of them urged women to continue their lawsuits. One Houston lawyer pursued more than 1,000 claims outside the global settlement.

In May of 1995, after a federal judge said the $4.2 billion that Dow Corning had committed to settle the claims would not be enough, the company filed for voluntary bankruptcy. Nobody ever proved that any implants harmed anyone but a business worth billions of dollars was bankrupted.[158]

In January of 1993, ABC Television's 20/20 public affairs show tracked some of the lawsuits that destroyed the multi-billion dollar Johns Manville Corporation and six other companies that produced asbestos or products that included asbestos.

The case is not a direct parallel to the Dow case, because there is evidence that asbestos can cause health problems and that some companies continued to produce or use asbestos after they knew of the problems.

The asbestos companies paid millions for their mistake but most of the money was collected by lawyers. Most of the 500,000 people who sued for damages got only a few thousand dollars each, but the lawyers who orchestrated the lawsuits got millions.

And some innocent companies were harmed. Keene Corp has more than 4,000 employees making hundreds of assorted products, but it made the mistake of buying a small company that had once made asbestos ceiling tiles.

The company cost $8 million to buy and the litigation over a product it was no longer making when Keene bought it cost more than $530 million. Plaintiffs got about $210 million of the settlement, and lawyers collected about $350 million.

Some lawyers do a lot of harm, and they are very well paid for it. One reporters survey found that several American lawyers make more than $10 million a year and that the best-paid lawyer in the U.S. received more than $90 million in fees in 1996.[159]

And we are all victims, because corporations of all kinds have to guard against assault by predatory lawyers. ABC News reporter John Stossel said lawsuits for 'slip and fall' accidents cost the city of New York about $200 million a year and the cost of lawsuits and insurance

[158] ABC News Special, The Trouble with Lawyers, Jan 2/96, host John Stossel. See also http://www.forbes.com/sites/henrymiller/2015/03/04/infuriating-titbits-about-silicone-breast-implants/#5a9dae0d18d6

[159] Ibid.

adds $100 to the price of the average football helmet, $500 to the price of a car and about $3,000 to the price of heart pacemaker.[160]

In May of 2017, a news article reported that the lawyers overseeing the re-distribution of Bernard Madoff's Ponzi-scheme profits to his victims had paid themselves $38.8 million but, after four years, had not returned any money to Madoff's investors.[161]

In an op-ed article in the April 20, 2011 edition of the *Washington Examiner*, Lawrence J. McQuillan and Hovannes Abramyan of the Pacific Research Institute calculate the cost of litigation — including annual damage awards, plaintiff attorneys' fees, defense costs, administrative costs and dead weight costs — from torts such as product liability cases, medical malpractice litigation and class action lawsuits — at $328 billion per year.

That's the direct cost but, the article says, the threat of litigation also changes behavior — and often in what the article described as "economically unproductive ways."

The article notes that fear of medical liability prompts doctors to engage in 'defensive medicine' — at a total cost of $124 billion a year. This is reflected in the high cost of medical insurance which, it estimates, adds 3.4 million Americans to the rolls of the uninsured. Because people without good medical care lose more time from work, the article estimates that this costs $39 billion per year in lost production.

The article suggests that legal costs also hamper innovation because companies spend money on legal defense rather than research and development, and they withhold or withdraw products from the market. It argues that more than 51,000 U.S. jobs have been lost due to asbestos-related bankruptcies alone and that employees at the bankrupted companies have lost $559 million in pension benefits. It suggests that side effects of tort cases cost $537 billion per year in addition to the direct cost of $328 billion for a total of over $865 billion per year. It also blames tort cases for an annual loss of $684 billion in shareholder value.

Civil law protects wealthy people and corporations from critics with 'SLAPP' suits — the acronym stands for "Strategic Lawsuit Against Public Participation" — that threaten to burden the victim with heavy legal costs. These are enabled by an early 17th century legal decision that if someone claims to have been libeled, it's up to the defendant to prove that they have not libeled their accuser, rather than the reverse. Typically, the complainant has lots of money and perhaps lawyers on staff, and the defendant can't afford a proper defense. Even if the suit has obviously invalid, the defendant will have to pay for a legal defense that he, she or they may not be able to afford.

I saw one example of legal bullying a few years ago when the chairwoman of a committee in the co-operative apartment building

160 Ibid.

161 Larson, Erik, *Bloomberg*, printed May 24/17 on page A1 of the *Toronto Star*

where I live sued some of her committee members. The chairwoman was a lawyer and at one point she told our building manager that the suit cost her nothing, because her lawyer managed it as a 'professional courtesy,' but it could have cost her committee members thousands of dollars.

Because the suit was so obviously unjust the co-operative hired a lawyer to defend the committee members — but that cost the co-operative, as I recall, $5,000. I know details of the case because I was on the co-operative's board of directors at the time.

Managers

Some of the most rapacious of the new takers are the 'professional managers' who now control many large corporations. Because the ownership of most big corporations is now spread among so many shareholders that no individual has a controlling interest, many of them have been taken over by hired managers in what author James Burnham called the 'managerial revolution.'[162]

He likened this takeover of the corporate decision-making process by managers to the takeover of government policy by appointed officials who remain in place no matter what party is elected and who pursue whatever course suits them.

North Americans have lost control of the private corporations in which they own shares. Most big businesses in the Western world are owned by shareholders, but no individual shareholder has much power and there is no check on the hired CEOs and directors. Where big blocks of stock exist, most are owned by unions or mutual funds which are themselves controlled by managers.

In 1932, a study by economist Gardiner Means and author Adolf Berle found that about half of all American corporations were controlled by their management, not their owners.[163] In 1963, a study by Robert J. Larner found that of the 200 largest corporations in the United States, 169 were controlled by their management rather than by their owners.[164]

Top executives of modern corporations don't own them and, as J.K. Galbraith found when he was deputy head of the U.S. Office of Price Administration and as management consultant Peter Drucker

[162] Burnham, James, *The Managerial Revolution*, John Day Co., NY, 1941, republished by Penguin in 1962.

[163] The study is reported in Means & Berle, *The Modern Corporation and Private Property*, published by Transaction Publishers, Edison, NJ and cited by J.K. Galbraith and Nicole Salinger in *Almost Everyone's Guide to Economics*, Houghton Mifflin, Boston, 1978, pg. 55

[164] Larner, Robert J., "Ownership and Control of the 200 Largest Nonfinancial Corporations, 1929 and 1963," *The American Economic Review*, vol. 56 No 4, Pt 1, (Sept 1966, pg. 777 et seq.

says in *Concept of the Corporation*, they don't run them,[165] but many CEOs collect tens of millions of dollars a year in wages and stock options.

A review of CEO compensation in Canada by economist Hugh Mackenzie for the Canadian Centre for Policy Alternatives shows that the average earnings of Canada's corporate top 100 increased by 178 percent between 1998 and 2015. In 2015, Mackenzie reports, Canada's 100 highest paid CEOs were paid an average of $9.5 million each — about 193 times the average industrial wage in Canada and more than 408 times the pay of someone who worked a full year at the average Canadian minimum wage of $11.18 per hour.[166]

Even when companies are losing money, the CEO's pay may increase. In one case, the CEO of a Canadian company whose second quarter profits fell by 50 percent collected more than $8,500,000 in 2015, and he left the company in July of 2016.[167]

A CEO's pay is set or approved by the board of directors, but the directors are appointed by the executives, who also decide what board members will be paid for their trouble. In some cases consultants are called in to determine a 'fair' rate of pay for the executives, but the consultants know that consultants who recommend high rates of pay are engaged more often, and paid more, than those who recommend low rates of pay.

Writing in 1978, John Kenneth Galbraith noted that one business executive who headed a company that was not doing very well was paid nearly $916,000 for a year's work but Cyrus Vance, doing the much more responsible and difficult job of being Secretary of State (and doing it very well, Galbraith says) earned $66,000.[168] In the modern world the President of the United States is paid $400,000 a year — a sum that vice presidents of many medium-sized corporations would sneer at.

Author and *Wall Street Journal* reporter Robert Frank says that in 1970, the average pay of American CEO's was about 40 times the pay of the average American worker. By 2007, the average pay of a CEO was 170 times the pay of the average worker.[169]

This despite evidence that high pay for top executives is counter-productive.

---

[165] Galbraith, John Kenneth, *The New Industrial State*, second ed, paperback, New American Library, 1979, pg.60. Footnote on pg. 80 cites Drucker, Peter, *Concept of the Corporation*, Transaction Publishers edition, 7th printing, 2008, Rutgers U, Piscataway, NJ., pg. 26. (First published by The John Day Co, NY, 1946.)

[166] Mackenzie, Hugh, *Throwing Money at the Problem 10 Years of Executive Compensation*, on line at https://www.policyalternatives.ca/sites/default/files/uploads/publications/National%20Office/2017/01/Throwing_Money_at_the_Problem_CEO_Pay.pdf

[167] Koplin, Francine, "CEO pay hit a historic high in 2015," *Toronto Star*, Jan 3/17, pg. GT6.

[168] Galbraith, John Kenneth and Salinger, Nicole, *Almost Everyone's Guide to Economics*, Houghton Mifflin, Boston, 1978, p112.

[169] Frank, Robert, *Richistan*, Crown Publishers, NY, 2007, pg. 44.

In an article on leadership in *Scientific American Mind*, magazine psychologists Stephen Reicher of the University of St. Andrews in Scotland, Alexander Haslam of the University of Exeter and Michael Platlow of the Australian National University in Canberra cite financier J.P. Morgan, who said the only common feature of the failing companies he worked with was a tendency to over-pay top executives.[170]

They also quote writer and professor Peter Drucker who wrote in *The Frontiers of Management*[171] that "Very high salaries at the top ... disrupt the team. They make .... people in the company see their own top management as adversaries rather than as colleagues .... And that quenches any willingness to say 'we' and to exert oneself except in one's own immediate self-interest."

In an experiment Reicher, Haslam and Platlow created work teams in some of which the leader was paid the same as his followers while in others he was paid twice or three times as much. They found that while high pay did not appear to increase the leader's efforts, high pay for the leader did appear to diminish the followers' efforts.[172]

Some CEOs are not actually expected to manage the companies they 'head.' Instead, they are hired as shills to boost the price of the stock and convince pension, mutual and hedge funds to buy it.

In this role they may provide real value for savvy stockholders who buy in before the price goes up and get out before the crash. In the never-never land of the financial industry, speculators who buy, sell and manipulate stock — and never produce anything of value — can make big money.

The United States is now technically bankrupt with a national debt — greater than the net worth of the country — that can probably never be repaid. In the summer of 2011, the country that calls itself the richest in the world nearly defaulted on its debts because a clique representing the wealthiest Americans refused to consider an economic package that would require rich Americans to pay higher taxes. This was about the time that Warren Buffett, said to be one of wealthiest men in the world, announced that his secretary was taxed at a higher rate than he was.

North Americans have lost control of the private corporations in which they own shares. As Burnham suggests, many of the paid managers who now run the United States are spiritual brothers of the bureaucrats who ran the U.S.S.R.

That's kind of humorous when you think about it — many executives who pretend to support an ethic of free market capitalism manage businesses they did not create and do not own, and they

---

170 Reicher, Stephen, Alexander Haslam and Michael Platlow, "The New Psychology of Leadership," *Scientific American Mind*, July 31/07, Vol. 18 Issue 4, pp. 22-29.

171 Drucker, Peter, *The Frontiers of Management*, Dutton, NY, 1996.

172 Reicher, Stephen, et. al, "New Psychology," pp. 22-29.

are more directly comparable to a Communist commissar than to a capitalist.

Some people still make fortunes by developing new ideas and new products but, for the past 50 years or so, many major fortunes have been made by people who do not actually create anything. Most Americans lost money in the crash of 2008 but, riding the market, two hedge fund managers made personal incomes of more than two billion dollars each.[173]

When The Baron conquered and occupied the craft village he outlawed the kind of banditry he had practiced in the past, but he enabled new and more effective methods.

Schools

Members of a herd generally think alike because they are often close relatives and, whether relatives or not, they are raised and live together. Artisans in a craft village might hold different sets of conventional wisdom but, whether they share the same craft or not, they all appreciate inventiveness and skilled workmanship.

The Baron's conquest of the village introduced a new group — soldiers who might have a common appreciation of the values of robbery and bullying but have little in common with the villagers. While it didn't matter much what the common people of the village believed or liked, it was important to The Baron and his cohorts that members of the conquering elite have common beliefs and values and that these beliefs and values be passed on to their children. One way to accomplish this was to invite the shamans and priests who supported The Baron to supervise and standardize the training of children of the elite. This worked, and even now churches run some of the world's best schools.

The development of schools has been a boon to mankind — I can write this and you can read it because we both learned to read and write — and they help to unify people by installing standardized conventional wisdom in all their students; but they have not been an un-mixed blessing. Many of the first schools served relatively small groups of people within a community and encouraged them to see themselves distinct from others.

Through most of history education has been divided into two streams — skills were taught by artisans to apprentices and 'book learning' was taught by priests or lay teachers to 'scholars,' many of whom were training to be priests. Because priests out-ranked artisans 'scholarly learning' — such as the study of dead languages, mythology, superstition and the misinformed beliefs of dead scholastics — out-ranked practical training.

[173] R Martin, Roger L. *Fixing the Game*, Harvard Business Review Press, Boston, 2011, pg. 32. Martin cites the 2008 Hedge Fund Rich List + *Alpha* magazine

It didn't matter much what the common people believed or knew but, as the industrial revolution began to demand more and more conformity and submission, it turned out that commoners raised with a degree of freedom were not all willing to submit to strangers they did not respect or to timetables that did not suit their convenience.

As Author Alvin Toffler explained in *The Third Wave:*

> The early mine, mill and factory owners of industrializing England discovered, as Andrew Ure wrote in 1835, that it was "nearly impossible to convert persons past the age of puberty, whether drawn from rural or from handicraft occupations, into useful factory hands." If young people could be prefitted to the industrial system, it would vastly ease the problems of industrial discipline later on. The result was another central structure of all Second Wave societies, mass education
>
> Built on the factory model, mass education taught basic reading, writing and arithmetic, a bit of history and other subjects. This was the "overt curriculum." But beneath it lay an invisible or "covert curriculum" that was far more basic. It consisted — and still does, in most industrial nations — of three courses: one in punctuality, one in obedience and one in rote, repetitive work. Factory labor demanded workers who showed up on time, especially assembly-line hands. It demanded workers who would take orders from a management hierarchy without questioning. And it demanded men and women prepared to slave away at machines or in offices, performing brutally repetitive operations.
>
> Thus from the mid-nineteenth century on, as the Second Wave cut across country after country, one found a relentless educational progression: children started school at a younger and younger age, the school year became longer and longer (in the United States it climbed 35 percent between 1878 and 1956) and the number of years of compulsory schooling irresistibly increased.[174]

Modern schools continue the program Toffler described, and they offer some extra courses that Toffler did not mention. One is that schools prepare children for life in a hierarchical world by teaching them to be helpless.

Psychologists around the world have studied the phenomenon they call 'learned helplessness' in dogs, cats, rats, cockroaches, goldfish and people since the 1950s, but schools have been teaching students to be helpless since the early days of mandatory schooling.

Experiments show that when any animal learns that it has no control over its environment, it stops trying to control that

[174] Toffler, Alvin, *The Third Wave*, Bantam, 7th printing, August/82, pg. 29. He cites Ure, Andrew, *The Philosophy of Manufactures*: or, *An exposition of the scientific, moral, and commercial economy of the factory system of Great Britain*, published by C. Knight, London, 1835.

environment. In a typical experiment, dogs are placed in a 'shuttle box' which has a low partition down the middle. The floor of either side can be electrified to give the dog a shock but, if only one side is electrified, the dog can avoid the shock by jumping the partition.

To condition a dog to feel helpless, experimenters electrify both sides of the box at the same time, so the animal cannot escape the shock by jumping the partition. After conditioning, the dog will cringe and whine and urinate when it is shocked, but it will not jump the partition.

For the second stage of a typical experiment the experimenters put a second, unconditioned, dog in the box and electrify only one side at a time. The new dog will jump the partition to avoid the shock but the conditioned dog will not. Even though it can see that the other dog avoids the shock, it will not jump the partition.

People react the same way. I don't know of any experiments in which people were given electric shocks, but a late 1970s study at a Veterans Administration hospital at Northport, New York, found that after a few weeks in hospital, patients seem to be conditioned to be helpless.

In her report of the experiment, psychologist Maritza Aminita Jonas noted that hospital patients have little or no control over their environment. "Choice of food and time of meals, television station, physician availability, standard procedures (such as temperature and blood pressure readings) not to mention the time and type of medical information concerning the individual, are all important factors over which the patient has little or no control."[175]

School children learn that they are helpless when they have to spend most of their day in a classroom, not talking to their friends and learning or pretending to learn things that do not interest them.

In people and animals, Jonas says, learned helplessness makes the individual passive, slows learning and reduces an individual's sexual and social functions.[176] These are all positive advantages to a eusocial society.

Schools also teach obedience and submission, and they support the development of a eusocial society by diminishing the influence of families.

This is a valuable function because if children learn everything from their parents, they will consider their parents to be the ultimate authority. If they learn in school, from teachers who may tell them their parents are wrong, they will learn that the teacher is an authority and — because teachers defer to official orders — that officialdom is the ultimate authority.

175 Jonas, Maritza Aminita, *Learned helplessness, the hospital setting as a conditioning paradigm*, abstract from a 1979 dissertation, published by University Microfilms International, Ann Arbor MI, 1983.

176 Ibid., p 16.

In the world of individuals, we respected and sometimes obeyed humans. The leader of a hunting party, for example, led because he knew the country and/or the quarry well and because most of the members of the party thought he was the most competent. Other hunters cooperated with him, rather than obeyed him.

Now we are expected to defer to and obey teachers, foremen, managers and others, most of who have been appointed by people we have never seen and who have never proved their competence to us.

In many cases we don't know whether our immediate superiors know how to do the job they supervise or not. In some cases we know that it does not matter how much or how little our immediate superiors know because they are just conduits for decisions made by others who have no direct contact with the situation and who may know very little about it.

Schools serve as babysitters, freeing parents to work rather than raise their families. Because others take care of their children, fathers can be made to feel that their jobs are more important than their families and even some mothers who can afford to stay home consider it more important to have a job than to raise a family.[177]

Schools destroy initiative by teaching children that the only acceptable activity is whatever the teacher tells them to do. Pupils are expected to act as they have been told to act and original ideas are liable to be punished. In a natural family a child will pay attention to whatever interests him or her at the time and will decide in his own mind whether to play or to build something or to sit and watch. Children in school learn to do as they are told, and that what they think or want is irrelevant.

Schools teach children to concentrate on things that are of no interest to them. To a healthy mind, a bee buzzing around a flower or a puppy playing with a ball are far more interesting and important than, say, a list of the kings of England or presidents of the United States. School teaches children that their own judgment is of no importance and that someone 'in authority' will tell them what to be interested in.

In many schools, children can't even go the bathroom without the teacher's permission. This is an important lesson for children who will grow up to be dominated by systems that will, eventually, control most of their waking hours and many of their dreams.

A corollary of learned helplessness is the phenomenon that educational theorist Ivan Illich called "the need to be taught." In *Deschooling Society*, he argued that, given the opportunity, most of us could learn most things on our own.[178]

[177] In most cases we rationalize that we work for money to support our families. This is generally true, but many of us work more to support our egos than to support our families.

[178] Illich, Ivan, *Deschooling Society*, Harper & Row Perennial, New York, 1972.

This appears to be true. Since the advent of computers, many adults have failed to keep up with the new skills required to use them but, as has often been recognized, children who have grown up with computers can often learn to use new programs on their own. Those of us who have learned 'the need to be taught,' have problems, partly because most computer programs do not include well-written instructions.

Above all, schools have taught most of us to believe in education. As the products of education themselves, teachers respect education and, as responsible teachers, they teach their students to respect it.

This can be a problem because schools teach only the conventional wisdom of their culture. Try to imagine, for example, an American school teaching the good points of Communism or a school in a communist country teaching the advantages of capitalism! I would not expect a Catholic school to offer a reasonable view of Protestantism or vice versa. Some American schools, even at the secondary and tertiary levels, teach a 'creationist' doctrine that rejects, and is rejected by, most of modern science. Within their communities, schools may be a stabilizing influence, but most of them offer a narrow point of view and some resist scientific and social progress.

Education should be respected, but not as much as some educators seem to think it should be. Let's remember, first of all, that the word 'education' usually refers to learning that is of little practical use. Learning that is of practical use is called 'training.'

An educated man may be able to read and write ancient Greek, but he can probably not build a house or fix a car. A trained man can build a house, fix a car, program a computer or do another of the thousands of jobs that keep our society working, but he probably can't read or write ancient Greek.

Perhaps the most important function of schools is that they standardize conventional wisdom. This is obviously a useful function — society is more stable if we all believe the same things — but it may also hinder progress. It's a bitter joke that you can measure the stature of a scientist by how long he is able to delay progress in his field — but in fact it's more than a joke. Some 'scholars' still respect Aristotle, some of whose wrong ideas were still taught and enforced by the Roman Catholic Church more than a thousand years after he died.

Much early education had little practical use but, because it was restricted, it offered anyone who could get it a chance to join the elite. Because educated people had esoteric 'knowledge' they were entitled to comfortable and/or well-paid positions, often in the service of noble or rulers.

Reading, writing and arithmetic were and still are useful, but these are just the basics and many artisans know them as well as an 'educated' man. In the Middle Ages a qualified mason, for example, had to be versed in advanced mathematics. A priest would have learned Latin, at least, probably Greek, both Roman and Greek history, astronomy, literature and other subjects which might be nice to know but are of limited use in day-to-day life.

Compare the educational accomplishments of a 21-year-old 'civilized' modern American or Canadian girl and an 18-year-old Native American girl from several hundred years ago.

The modern girl may have a university education but, unless she has a master's degree or better, the Native girl is better educated. Even if the modern girl has a university degree, the Native girl probably has more useful knowledge and skills.

Consider her accomplishments. Native North Americans used at least 1,500 different plants for food and hundreds more for other purposes. Not all tribes knew all plants, of course, but by the time a girl was considered fit to marry she would be able to identify hundreds of plants and know where to find or how to grow them and how to use all the different parts.

That itself is a lot of learning. Most plants have multiple uses. Consider, for example, the common bulrush. In spring you can eat the stalk, raw, or the head raw or cooked in batter. In fall you can break the head up into a kind of flour, which can be baked into cakes, and at any time of year you can dry the roots, grind them up and use the powder to make a drink something like coffee.

In addition to the plants a Native girl would be expected to know almost all the animals that live in her area. She would know where and how they live, how they can be hunted or trapped and their anatomy in detail. Given a dead animal, she could skin it and tan the hide, butcher the animal and preserve or cook the meat, and make thread or string from the tendons, household tools from the bones and so forth.

She would also know how to make all the clothes she, her husband and their children would require, how to make several different types of shelter, how to braid ropes, bowstrings and other fiber products, how to weave baskets and perhaps blankets and how to make most of the tools, weapons and implements people need to live in comfort. Native girls around the world would have comparable training and skills.

Most of the 'civilized' girl's knowledge, on the other hand, is superficial. She knows of modern machines but she probably does not know how they work. She may know the names of some dead kings and Greek philosophers, but the Native North American girl is just as likely to know the legends of her tribe.

The civilized girl may speak a second language, but the Native girl may also speak a second language. The modern girl may know how to use or even to program a computer but she probably does not know how to find or prepare her own food, clothing or shelter.

The Native girl's knowledge is different from the modern girl's, but is more useful and more extensive.

The ultimate test is the ease of conversion from one culture to another and primitive men and women around the world can easily absorb the knowledge required to live in our culture. When they do have problems, it is usually with social or emotional issues. Few people from our culture, on the other hand, can learn to take care of themselves without the help of civilization.

Many teachers claim that education teaches you to think, but that's open to question. Most students know they can get good marks by showing how one can rationalize their teachers' conclusions from their teachers' premises, and that it is not wise to question either the premises or the conclusions. I know some people who were so well educated in school that they do not see the need to learn anything in later life.

We might compare the training, responsibilities, pay and prestige of medical doctors and automobile mechanics. A mechanic's training is more practical than a medical doctor's, and it takes almost as long.

The biggest difference is that doctors work with two basic models which have not changed for thousands of years, and those models can tell you where they hurt and will repair themselves if given half a chance. Mechanics work with dozens of models that change every year, that can't tell you what's wrong, and that will not repair themselves.

Some doctors accomplish miracles but some are quacks. George Washington died of a minor throat infection that his doctors treated by bleeding him of nearly three quarters of his blood, putting beetles on his legs to raise blisters, and burning his neck.[179]

Through much of European history, many medical practitioners did more harm than good and, in modern times, we know that some totally un-trained people have been able to masquerade as qualified doctors and get away with it. Some mechanics probably fake qualifications too, but it's easier for a doctor than for a mechanic to fake success.

If a doctor gives you something that makes you feel worse for a while you will feel better as the effects wear off and you may think you are cured. We find it hard to tell because humans are almost infinitely adaptable, and we can get used to discomfort.

But it's hard for a mechanic to fake anything, because our cars either work or they don't. A mechanic may pad the bill but if my car does not start I will know it, if it burns too much gas I will know it and so forth. A lot of people take better care of their cars than they do of their bodies, possibly because we watch our cars from the outside and we are more aware of their performance.

We can also compare truck drivers and airline pilots. The pilot takes off and lands only from airports designed for his plane. Mechanics check his plane before every flight and he has teams of

[179]http://www.medicalbag.com/what-killed-em/george-washington/article/486644/

supervisors and assistants to watch over his work and warn him of problems. In a ten-hour flight, he has to be very careful for ten or fifteen minutes around take-off and landing, but he can switch on the autopilot and relax or even sleep most of the time.

A truck driver has no help on the road and he is in constant danger of being cut off or sideswiped. He has to check his own truck before each run and, on a ten-hour drive, he can't afford to let his concentration slip for a minute.

I don't say that an automobile mechanic deserves more respect and more pay than a medical doctor, or that a truck driver should get more than an airline pilot, but I do suggest that our attitudes toward all of them could use some adjustment.

If you really want to see what jobs are worth, consider what happens when people don't do them.

When doctors go on strike people worry, but when garbage collectors go on strike we have a serious health problem. If airline pilots strike, tourists' vacations are interrupted but if truck drivers strike, our cities run out of food.

But theoretical learning is still considered better than practical knowledge, and it has been since before the dawn of civilization.

At some point in pre-history men and women who developed special skills in the technology of the day — arrow making or flint knapping or basket weaving or whatever — were able to drop out of the daily grind and specialize in the work they did best. Most of them taught their trade to their children as they worked, and sometimes to other children who were interested.

Possibly about the same time, other people began talking about gods. The artisans obviously contributed more to society than the priests, but the priests had more prestige because they interpreted the will of the gods. In most cultures it seems that what the gods wanted most was for the people to support the priests in luxury.

Some hunters became robbers and soldiers and they held most of the real power, but priests had power too. In many cultures they studied the stars and developed a calendar to tell farmers when to plant their crops. In Egypt, Peru and other areas they developed surveying techniques and mathematics, so they could supervise the construction of canals for irrigation.

Priests told fortunes, learned to read and write, treated sick people and interpreted laws, And they also taught students — originally priests-in-training and later the sons of the wealthy. Some of the illegitimate sons of nobles were taught enough to serve the priests as lay brothers or scribes.

But all of them were privileged, and through most of history students were an exclusive group. Partly because admission to school was restricted, education became a ticket to the elite. If a commoner could get an education he could probably get a soft job in the service of some noble. Because educated people shared esoteric knowledge with other educated people they were entitled to soft and/or well-paid positions in which they did not have to do much work.

But beyond reading and writing, education did not have to include any practical knowledge or skills. Civic administrators in ancient China were chosen by an exam in which candidates were required to write a poem. Their work was judged on the quality of the poem and of the calligraphy, both of which were important to administrators who were themselves chosen for their poetry and calligraphy.[180]

Educational qualifications also filtered senior administrators in the British Empire, but the standards were different. In Britain an applicant would be hired for a good job if he came from an exclusive school and for a lesser job if he came from a lesser school.

The British system made selection easy because relatively few people went to exclusive schools, and it worked because senior administrators don't have to know or understand much anyway. (Remember von Holst's experiment? The fish with the lobotomy became the leader.) If actual knowledge or judgment is required, senior administrators have underlings to provide it.

Some modern schools provide technical training but, above all, a university degree is proof that the graduate has been accepted into the lower levels of the elite and has not been a serious rebel. In this it performs a sorting function because people who have not been able to afford or have not been accepted by a university need not be considered for membership in the elite. If they do not complete their education it may be because they are too independent or self-motivated to put up with the restrictions of school and would therefore not make good members of the elite. Most people who have chosen or been chosen for 'training' rather than 'education' in their youth are liable to be type-cast for life, and have little chance of being considered for a 'management' job.

Education also teaches us to accept information only from approved sources. Most of the ideas in this book are my own, but I use lots of quotes and footnotes to prove that they are compatible with 'authorized' sources. Even so, many scholars will refuse to consider my work because it conflicts with conventional wisdom.

School Sports

In ancient Greece Plato's *Academy* stressed sports and military training, partly because Plato and most of his students owned slaves and they had to be prepared to crush a rebellion. Most modern students are destined to be wage slaves rather than slave owners, but school sports and other extracurricular activities still serve a purpose,

[180] Calligraphy would be important in a culture in which laws and decisions were written by hand, but a good administrator who was a poor calligrapher could always hire a better calligrapher. A good calligrapher who was a poor administrator would not be likely to hire a better administrator.

because students who join school teams or a band or other school activity learn to work as a team rather than as individuals.

In North America, where we think of ourselves as 'individuals,' some of our most popular sports leave very little room for individual action.

Compare, for example, American or Canadian football with the world football game that we call 'soccer.' In American or Canadian football an attack is usually an organized 'play' which is often a traditional maneuver developed years ago, and practiced and performed by rote. In a game the 'play' for each situation is chosen by one person — nominally the quarterback but sometimes the coach — and each player does exactly what he has rehearsed.

Soccer players play for a team too, but each man is expected to think for himself. Each player has a 'position' to cover but once the game starts, there is no one to tell him how to cover it.

I don't deny the value of 'team spirit' because cooperators always win over individuals in the long run. Still, for the long term welfare of humanity I think it's important that we know who we are cooperating with; and the 'team spirit' that most sports encourage never questions the choice of team.

Advanced Education

Advanced education for the masses is a recent development and most people see it as a benefit, but that is open to question.

Advanced education is useful partly because it gets people over the dangerous period in which they are open to new ideas before they enter the work force and long before they are in any position to make important decisions.

In his *General Theory of Employment, Interest and Money*, economist John Maynard Keynes wrote: "There are not many who are influenced by new ideas or theories after they are twenty five or thirty years of age, so that the ideas which civil servants and politicians and even agitators apply to current events are not likely to be the newest."[181]

University education has the advantage that it takes young people out of their homes and exposes them to an approved version of conventional wisdom without any parental influence that might counter it. At mass-production universities, whole generations can be programmed to think along the same lines. This is an obvious benefit to a eusocial society.

[181] This is said to be Keynes' most famous quote, but it's also one of the hardest to track down. It's in chapter 24 of the 1936 and 1947 editions of the *General Theory* and on the last page of the 1939 edition but not in later editions. It's possible that Keynes changed his mind but I think it more likely that his publisher reacted to the yowls of outraged academics, and considered the possible effect on the acceptance of Keynes' work as course texts. If you google it, you'll get lots of references.

Advanced education also helps to create and maintain a culture of debt. Many university students have to borrow money to pay for their education and living expenses and they are deep in debt before they start work.

This ties people into society because people with debts must take whatever job is offered and put up with whatever it demands. Once they learn to live in debt, they will accept it as normal.

This works for society because people who are in debt can't afford to rebel. They are tied to a treadmill and, because they are conditioned to it, they are glad to be there.

University education also helps to put a brake on both technical progress and social change. Most universities pretend to search for new ideas but the fact is that most university professors and lecturers have completed their own education, and many of them see no need for new ideas.

Priests, scholars and rulers tend to resist new ideas. Ancient Egypt and China developed considerable technology in their formative years but after the priests and pharaohs and emperors gained power, technical progress stopped for thousands of years. Chinese ships traded over most of the world before the European voyages of discovery but, nearly 70 years before Columbus sailed from Spain, an Imperial edict banned the construction and use of ocean-going ships and Chinese colonies in Africa, Australia, New Zealand and South America were abandoned. At one point it was illegal for a Chinese to learn a foreign language or to teach Chinese to a foreigner.[182]

Many of the technical developments we enjoy today were made by practical men like Thomas Edison, George Westinghouse, Henry Ford and Wilbur and Orville Wright. Edison had only a few weeks of 'education' in a one-room schoolhouse, Ford had eight years of schooling before he went to work in a machine shop, Westinghouse lost interest in schooling after one term at college and the Wright brothers were bicycle repairmen. Elisha Otis, inventor of the emergency brake that prevents an elevator from falling if the cable breaks, was a mechanic who worked as a wagon driver then built his own grist mill, converted it to a sawmill, built wagons and carriages and worked at an assortment of factory jobs before he began making elevators. John Deere was a blacksmith and Clessie Cummins was a chauffeur. All of these pioneers might be considered 'unemployable' in the modern world and if they did find jobs they might not be able to keep them because, like other great innovators, they had more initiative and independence than modern employers like to see in the kind of employees that get their hands dirty.

In the 1970s and 80s, computer technology was changing overnight and many of the most important advances were made by inventors who

[182] Menzies, Gavin, *1421, The year China discovered America*, William Morrow, NY, pg. 53, 55. The book was first published by Transworld Publishers in Ealing, England.

dropped out of school to develop their own ideas. Some wags suggest that multi-billionaire Bill Gates might have become a millionaire if he had not dropped out of university. Billionaire Sir Richard Branson — founder of Virgin Records, Virgin Airlines, Virgin Cellular and other companies — dropped out of school at 15. Andrew Carnegie dropped out at 13 and Frank Stronach, billionaire founder of Magna International, dropped out of school to apprentice as a tool and die maker when he was 14 years old.

Advertisers and marketers continue to trumpet 'new' products, but most of the real developments of the modern world are in communications, record keeping and military weapons. We get a lot of ballyhoo about 'improvements' in consumer goods but most of the changes are either cosmetic or made to reduce the cost or increase the price or the sale of goods. Changes that increase the utility, efficiency or durability of goods are sometimes deliberately avoided.

Americans learned this in the 1970s when they discovered that the cars they thought were the best is the world were, in fact, below world standard in quality and performance. American cars are now much better, but many of the old brands are gone, one of the three surviving auto makers is now foreign-owned and even many Americans now consider many European, Japanese and Korean cars to be better than competing American products.

Governments and private entrepreneurs pour hundreds of billions of dollars into astronomy and the exploration of space, even though space flight with chemical rockets is so horribly expensive (and so harmful to the environment) that it will never be practical, and we have lots of room for exploration in our oceans.

Some private companies and government agencies spend billions on climate change — some to prove that it is happening and some to prove that it is not — without ever thinking that it has happened before and will happen again. Granted that the use of fossil fuels may exacerbate it, there is no way we could stop it. We could prepare for it and perhaps save tens or hundreds of millions of lives — perhaps even allow the system we call 'civilization' to survive — but the takers who dominate our culture don't see the benefit in that.

Danish theoretical physicist Per Bak offers an example of hidebound scholasticism. In *How Nature Works, the science of self-organized criticality*, he wrote,

> I have the deepest respect for the type of science where you put on your rubber boots and walk out into the field to collect data about specific events. Such science provides the bread and butter for all scientific enterprise. I just wish there was a more open-minded attitude toward attempts to view things in a larger context.
>
> I once raised this issue among a group, not of geophysicists, but of cosmologists at a high table dinner at the Churchill College in

> Cambridge. "Why is it that you guys are so conservative in your views, in the face of the almost complete lack of understanding of what is going on in your field?" I asked.
>
> The answer was as simple as it was surprising. "If we don't accept some common picture of the universe, however unsupported by the facts, there would be nothing to bind us together as a scientific community. Since it is unlikely that any picture that we use will be falsified in our lifetime, one theory is as good as any other." The explanation was social, not scientific.[183]

In some areas, it doesn't matter whether the 'science' is valid or not, as long as all the 'scientists' agree. That's no surprise because, like other humans, many scientists are herd animals and herd animals believe what the herd believes.

Still, having vented my spleen about advanced education, I don't pretend that we could maintain our present level of technology without it. In many fields we are past the stage where a natural genius like Edison can make significant advances with no education at all.

But when students are taught to think in conventional ways, few of them will break free to develop new ideas. As humanity has several times in the past, we have probably reached a plateau of technical progress.

It's hard to imagine how we might maintain our technological culture without a system of advanced education but, hundreds of years ago, the cathedrals that we still see as technological marvels were built by masons who learned by apprenticeship on the job while scholars studied dead languages and misconceptions about the nature of the universe. In more recent times most of the major advances — including the steam engine, cars, airplanes, elevators and computers — were developed by practical men who had little formal training in the fields they developed.

I do not insist that we could get along without advanced education, but I do argue that we would be better off without learned helplessness and the need to be taught.

The Einstellung Effect

All levels of education contribute to the phenomenon that psychologists call the 'Einstellung effect,' which is a close relative of conventional wisdom. The term refers to the common practice of trying to solve a problem in a way that has been learned, even when a simpler solution is obvious.

Psychologist Abraham Luchins demonstrated this in 1942, with his famous *water jar experiment* — in which subjects were given three

[183] Bak, Per, *How nature works, the science of self-organized criticality*, Copernicus, Springer-Verlag, NY, 1996, pg. 86.

water jars, each with a different capacity, and asked to measure out a specific volume of water different from the capacity of any of the jars.

In one run, for example, the subject may have one jar that will hold 127 fluid ounces, one that will hold 21 fluid ounces and one that will hold 3 fluid ounces — call them jars A, B and C — and be required to measure out 100 fluid ounces.

The solution in this case is to fill A and then, from A, fill B once and C twice. 127 fluid ounces minus 21 fluid ounces minus 3 fluid ounces minus 3 fluid ounces equals 100 fl.

In Luchin's experiment, one group was given five practice problems followed by four test problems, and a second group was given only the four test problems. All five practice problems and some of the test problems had the same solution, A minus B minus 2C, but at least one of the test problems could be solved with a simpler procedure.

If the subjects were required to measure out 18 fluid ounces of water with jars that would hold 15, 39 and 3 fluid ounces, for example, they could fill the 15 fluid ounce and the 3 fluid ounce jar, and dump them both into the 39 fluid ounce jar — but many of the subjects who had been given the practice problems used the more complex solution, 39 fluid ounces minus 15 fluid ounces minus 3 fluid ounces minus 3 fluid ounces.[184]

If we learn to solve problems in school, we learn the solutions that our teachers learned — which may be the solutions that their teachers learned. Some university students still study the 'facts" of the world's 'creation' and believe that humans, and other animals, are products of 'intelligent design.' As noted earlier, most of the technical advances of the modern world were made by men with little formal education.

Traditional beliefs are a problem for humanity because they hinder or prevent the development of new ideas, but they are a benefit to the established order, for exactly the same reason.

Behavioral economist Richard Thaler offers one example of this problem from his own experience. His field is distinct from conventional economics because conventional economists predict the behavior of idealized but non-existent rational consumers under imaginary conditions, while behavioral economics studies the behavior of real people in real conditions. In a question and answer session after Thaler had outlined his recent work at a conference, one well-known economist rose to speak.

"If I take what you are saying seriously," he said, "what am I supposed to do? My skill is knowing how to solve optimization problems."[185]

That well-paid professional had only one marketable skill — his ability to apply conventional methods to solve conventional problems in conventional ways — and when and if the world adopts the more-realistic science of behavioral economics he, and thousands more

184 https://en.wikipedia.org/wiki/Einstellung_effect

185 Thaler, Richard H., *Misbehaving*, W.W. Norton, NY, 2015, pg. 43.

comfortably ensconced 'professionals,' will have to be re-classified as untrained and unskilled.

Plans

Education also encourages us to make plans. This is not a new development because most pack hunters make plans and The Baron was, in many respects, the leader of a hunting pack.

Eusocial insects have neither leaders nor plans. They have queens, but the queen's function is to lay eggs and she serves the colony or hive, rather than rules it. Worker ants, termites and honeybees forage for food while others tend the young and repair and maintain the nest but while individual insects may shift from one job to another, there is no central command. That sounds like a recipe for chaos but, somehow, every job that needs to be done is done, and decisions are made.

In one series of experiments in which a swarm of bees were forced to choose between nine possible nesting sites, they chose the one that was best suited to their needs. The decision seemed to have been made by about fifteen bees that left the swarm to explore, but that had no apparent authority.[186]

Swarm intelligence works for insects and for some human hunting and gathering groups, but many human communities have a ruler whose word is law. We can only guess how this evolved but we know that every successful army in history had a strong commander who did not share his authority, and ultimately, most of humanity is controlled by armies. In the 'free world' we believe that our governments rule by our consent but, with or without consent, governments rule because they control armies to defend them from external enemies and police forces to enforce their will against their own citizens.

On one level, rule by governments seems to work, but that may be because we see only a short term view.

In the past 10,000 years countless kingdoms and empires have formed and collapsed, but none of them lasted as long as the culture of the !Kung bushmen, which seems to have worked for more than 60,000 years, while that of the Australian Aborigines lasted for at least 35,000. Compared to them, the empires of Greece, Rome and even China were ephemeral.

We think of the !Kung and the Australian Aborigines as 'dead end cultures' and, if they had ever met them, no doubt the Babylonians, Hittites, ancient Egyptians, Romans and other long-lost 'civilized' people would have thought the same. I guess it depends on your definition of 'dead end.' The fact is that no complex culture ever lasted for a tenth as long as the !Kung and, given that complex cultures

[186] This experiment is described in the article "Swarm Theory," by Peter Miller, in the July 2007 issue of *National Geographic* magazine.

threaten to make the world uninhabitable, it does not seem likely that any will.

Most of the hunter gatherers of the world have been killed or enslaved by armies as 'civilization' — the word we use to describe cultures in which an elite is supported by slaves or peasants — spread but, if they are not wiped out by armies, simple cultures last a long time. We can only guess at the reason for the failure of kingdoms and empires and the survival of cultures without them, but one difference stands out. Rulers and their hangers-on live very comfortable lives and, because change might reduce that comfort, they may be slow to adapt to new conditions. In cultures where all are equal, there is no need to fear change and no one to oppose it.

It's also interesting that where we have rulers we have competition but where there are no rulers we can hope for cooperation. As noted earlier, Axelrod's computer tournament of *Prisoners' Dilemma* proved that in the long run the 'good guys' — AKA cooperators — always win and that 'bad guys' — who try to take advantage of others — lose.

Rulers make plans — in many cases, it is because they made plans that they became rulers. That's another problem, because plans can lead both planners and their followers astray.

I don't say that all plans are bad, or that they are unique to humans, but most plans work only in the short term and often they work better for the planners than for the people who are expected to carry them out. Human rulers tend to make long-term plans, and that's a problem because we can't foresee the future. Many long-term plans are made and implemented to solve the problems of the past or that will be past before the plans are completed.

Author Jane Jacobs outlined the failure of the Shah's plan to industrialize Iran in her landmark *Cities and the Wealth of Nations*.[187] In 1975 he decided to build a factory to produce large helicopters in the city of Isfahan which, at the time, had an economy that was more mediaeval than modern.

He hired an American company to design the helicopter and build the factory, and the American company subcontracted much of the work of building the factory but did not, in the end, design a helicopter. The project consumed most of Iran's huge oil income and savings, drove the country into debt and, when food and other prices began to rise, was a major cause of the revolution that deposed the Shah and made Iran and the United States bitter enemies.

Do you think Richard Nixon's plan for a 'war on drugs' worked out the way he expected? Did George W. Bush's invasion of Iraq establish peace and stability in the Middle East? How did the Vietnam war or Lyndon Johnson's 'War on Poverty' work out?

One problem is that plans made by rulers are often implemented by subordinates who may not be smart enough to see when a plan is

[187] Jacobs, Jane, *Cities and the Wealth of Nations*, Random House, Vintage Books, NY, 1984 (1985 printing), pg. 135-7.

failing and/or might not have enough nerve or authority to change or abandon a plan that is failing.

Even a senior bureaucrat or politician may not be able to change or abandon his own plan. If he has made it public and been elected or promoted on the strength of it, he can't afford to admit either that his plan is a failure or that he is not able to carry it out and that, either way, he does not deserve his position.

We don't hear much about the failure of plans, but that's at least partly because few people see the overall effect of a large-scale plan and it's often easier for the people in charge to hide a failure than to fix it. Computer programmers know this problem, and they have a standard way of coping with it. When they develop a new program they first write a 'quick and dirty' version of it, which they hand out as a 'beta' for testing. After the beta-testers find all the problems and mistakes the programmer writes the final version — and this may be two or three steps down the line.

Contrast that with politicians and other bigwigs who announce their plans to the public, then expect lesser lights to carry them out. Because it would harm the bigwig's reputation to find that the plan was un-workable the temptation — and often the procedure — of the lesser light is to carry the plan out as specified and fudge reports to make it look as though it had worked.

The government of the Canadian province of Ontario still trumpets the 'success' of the wind-powered electric generators it promoted even though most of the power they generate is sold, at a loss, in the United States. The heirs of the administration that bought the wind-powered generators also sold much of the publicly-owned power company to private financiers, and power rates in the province multiplied several times over the next few years. As this is written, the same political party is trying to save itself with a scheme that will stretch payments for some projects over a much longer time — and cost tens of billions of dollars more in interest charges.

In a brilliant book the late Jane Jacobs listed some of the disasters of city planning in New York and Chicago, where planners destroyed working neighborhoods and replaced them with 'settlements' and other developments that became breeding grounds for crime and violence. She wrote of the hatred tenants held for a lawn that was part of the redevelopment of a part of East Harlem, in New York.

A social worker frequently at the project was astonished by how often the subject of the lawn came up, usually gratuitously as far as she could see, and how much the tenants despised it and urged that it be done away with. When she asked why, the usual answer was, "What good is it?" or "Who wants it?" Finally one day a tenant more articulate than the others made this pronouncement: "Nobody cared what we wanted when they built this place. They threw our houses down and pushed us here and pushed our friends somewhere else. We don't have a place around here to get a cup of coffee or a newspaper even, or borrow fifty cents. Nobody cared what we need. But the big men come and look at that grass and say, 'Isn't it wonderful! Now the poor have everything!"[188]

In the city of Toronto, where I live, a neighborhood was demolished in the 1950s to build the infamous Regent Park development which has, thankfully, since been mostly torn down since then and replaced. In all cases, the problem is that the kind of layout that looks good to planners is too often ideal for the breeding and support of crime.

The Eighteenth Amendment to the Constitution banned the sale of alcohol in 1919, but the mistake was so obvious that Prohibition was repealed 1933. The 'War on Drugs' lasted more than 40 years after it was declared by President Richard Nixon in 1971, even though was obvious that it was a ghastly mistake. Even now, while some individual states are backing away from it, the Federal government has tens of thousands of employees and hangers-on who depend on it and billions of dollars invested in police equipment and prisons, a private prison industry that depends on it and manufacturers who depend on the slave labor they rent from private prisons.

Admirers of the Roman Empire like to say that Romans spread civilization over the lands they conquered. They also spread top-down rule and slavery. Many American politicians tell us that the United States is spreading democracy around the world. In fact the U.S. has helped replace elected governments in Iran, Argentina, Chile and other countries with dictators.

This is not a call to anarchy because, one way or another, we are now adapted to living under rulers, and we probably could not live without them — but it does give us something to think about.

For an objective look at long-term plans, let's review the work of some of the great planners of history.

Plato planned *The Republic* and people who have never read the book think it would have been a nice place to live, but Plato's ideal was a dictatorship supported by slaves and ruled by a 'philosopher king.' He believed that all women should be the common property of all men, that some men were born to be slaves, and that no one — slave or 'free' — should have any freedom. Consider this passage:

> The greatest principle of all is that nobody, whether male or female, should be without a leader. Nor should the mind of anybody be habituated to letting him do anything at all on his own initiative; neither out of zeal, or even playfully. But in war, and in the midst of peace — to this leader he shall direct his eye and shall follow him faithfully. And even in the smallest matter he should stand under leadership. For example, he should get up, move, or wash, or take his meals ... only if he has been told to do so. In a word, he should teach

[188] Jacobs, Jane, *The Death and Life of Great American Cities*, Random House, NY, 1961, Modern Library, NY, 1969. In my paperback copy, from Vintage Books, NY, p15. Jacobs lists other failed developments on pp 42-45.

> his soul, by long habit, never to dream of acting independently, and to become incapable of it.
>
> In this way the life of all will be spent in total community. There is no law, nor will there ever be one, which is superior to this, or better or more effective in ensuring salvation and victory in war. And in times of peace, from the earliest childhood on should it be fostered — this habit of ruling others and of being ruled by others. And every trace of anarchy should be utterly eradicated from all the life of all the men, and even of the wild beasts which are subject to men.[189]

Our word 'Utopia' comes from the book *Utopia* by St. Thomas More, who fictionalized his plan for a perfect society. His Utopia was an island, ruled by a hierarchy, in which citizens needed a pass to travel more than a mile from their home city. If they were caught without one, they could be sentenced to slavery. Adultery was also punished by slavery, and if the injured party wanted to forgive and continue the relationship, he or she had to join the guilty in slavery.

As the island becomes over-crowded, Utopians set up colonies on the mainland — pushing the natives aside and taking their land.[190]

That's Utopia, all right — for the rulers.

Edward Bellamy dreamed up a different 'perfect' society in *Looking Backward*, 2000–1887.[191] In the United States of the year 2000, as Bellamy envisioned it, citizens would have to work 24 years at an assigned job and anyone who refuses his assigned work is thrown into a solitary prison cell and fed only bread and water. Meals are cooked in a central dining hall and eaten in private rooms in the central hall, but people listen to music, by telephone, in private rooms in their own houses.

In Bellamy's ideal world, "young men are taught habits of obedience, subordination and devotion to duty."[192] Not as bad as More's Utopia, but certainly not free.

Many planned societies have worked well enough for the planners, and for the planners' lifetimes, but the first thing most planners plan is a hierarchy.

At this stage, it may be too late for humanity to try to live without plans, but I like the traditional Japanese approach, as outlined by Prof. Tadeo Umaseo in his keynote speech to the 1973 International Congress of Societies of Industrial Design in Kyoto. Too much clarity of purpose creates confusion, he explained, and the traditional

[189] Translated and quoted by Popper, K.R. in *The Open Society and its Enemies*, Princeton University Press, Princeton NJ, 5th ed, 1966, pg. 102.

[190] More, Thomas, (Sir, St.) *Utopia*, first published in 1515, I cite a modern copy by Broadview Press of Peterborough, ON, Buffalo NY and London, UK, 2010.

[191] Bellamy, Edward, *Looking Backward, 2000-1887*, first published in 1888. I cite The Belknap Press of Harvard University Press, 1967 publication, edited by John L Thomas.

[192] Ibid., p 171.

Japanese approach to development is to watch developments as they occur and adjust policy to accommodate them.

Umaseo was talking about Japan, but the best example of that attitude that I know of (if it is true) is American. There is a story that when former General Dwight Eisenhower was president of Columbia University, the grounds-keepers complained about students who walked across the lawns rather than follow the paved paths they had laid out.

Eisenhower's solution — replace the paved paths with grass and pave the routes the students walk on.

For a ruler raised with the idea that rulers plan and rule, a policy of guided drift would be a challenge; but it might be one worth facing. As an alternative we might consider the policy (apocryphal, but it sounds good) of the city state in which rulers held office for a fixed term. At the end of his term the retiring ruler was led to a gibbet, a noose was fitted around his neck and he was given his pay in cash.

Then they held the trial. Now there's a plan I could support.

The Persuaders

The Baron planned the conquest and occupation of the village, but he is not a one-man show. He has soldiers to threaten the villagers and officers to manage the soldiers, but threats of violence are not enough for complete control. He also has some agents — villagers and others who are loyal to him rather than to their neighbors — and this group may include some priests.

The Baron's rule will be more secure if the villagers forget that he took over by force and believe that the great god Gumball chose him to rule and that Gumball wants everyone to obey him. Sooner or later he will learn that the best way to get this idea installed in the villagers is to elevate the great god Gumball by honoring his priests and giving them the right to demand support from the villagers as a reward for teaching the true religion. By honoring the priests, The Baron establishes them as alphas and, because they are alphas, many people will believe what they say.

We can only guess how the first belief in supernatural powers got started, but some people assume that men must have wondered about how the world began and about natural phenomena like night and day, summer and winter, rain and snow, etc. Most cultures believe in gods or spirits of some kind and most have 'creation myths.' Even among hunting and gathering bands some shamans claim special power to interpret the will of the gods and to intercede with them, and some shamans were able to make other members of the band fear their power.

Religions and rulers form metasystems. If a religion backs the rulers of the day most rulers will back the religion — sometimes giving it the right to demand tithes from worshippers. Religion can be a unifying force and it may have been a benefit to the people of the village, but it has also been a benefit to rulers because it justified their rule — and to priests because the priests of a wealthy and powerful church will themselves be wealthy and powerful.

Through the centuries priests have initiated and managed many of the most expensive construction projects of history — including three of the 'seven wonders' of the ancient world and others that would have enlarged the list if the scholars who compiled it had known about them.

The people of Egypt did not need pyramids but the religion needed physical symbols of the power and importance of the priests, the kings and the gods. The people of Olympia did not need a statue of Zeus, but the one they built was another of the seven wonders of the ancient world. The people of Ephesus did not need a Temple of Artemis, but that was yet another of the seven wonders.

In the Americas the great Pyramid of the Sun in Teotihuacan, the temple pyramids of the Maya, the mounds of Cahokia in the United States and the temple of Sacsayhuaman in Peru were apparently built to serve religions. In Cambodia the ancient temples of Angkor-Wat are still magnificent.

Between AD 1050 and 1350, the people of France built 80 cathedrals, 500 large churches and thousands of parish churches. Encyclopaedia Britannica says the cathedrals and churches of France required more quarried stone than all the palaces, temples and pyramids of ancient Egypt.[193]

Temples are of no practical use to people but their construction is useful to rulers and priests because, to build a big temple, people must work in large gangs under the direction of overseers for a project that is of no direct benefit to themselves. This is a step toward eusociality, and gangs of men working together are an important feature of civilization. In a society without mass media, temples and churches provide a meeting place in which people can be told what to believe and do.

The construction of temples also taught priests and their delegates how to control large teams of workers. Archaeologists used to think Egyptian pyramids were built by slaves, but many now believe they were built by the 'voluntary' labor of 'free' citizens. That may be, but if 'free' citizens 'volunteer' that much labor, they are obviously dominated by the priests and kings who organize it.

Priests may have developed reading, writing and mathematics and they probably developed astronomy. All these are useful to humanity but they also make central control more practical. Priests in some areas produced calendars to tell farmers when to plant their crops and in others they supervised and managed irrigation systems. These

[193] https://www.britannica.com/technology/building-construction/Romanesque-and-Gothic

are all of use to humanity, but managing them reinforces the power structure.

At this point we have to remember that we are talking about functions, not people. Many priests have been devoted to the service of their gods and of mankind and some have been responsible for great accomplishments; but even priests who serve mankind must also serve their religion and the needs of the religion may not have much in common with the needs of either gods or men.

Mayan priests taught that the gods would be pleased if a teen-aged girl was weighted with gold jewelry and thrown into a well to drown. The Inca were taught that it would please their gods if young children were drugged and left on mountain-tops to freeze, and Aztec priests tore the hearts out of living victims.

In ancient times Jewish priests taught that it was virtuous to kill Canaanites. In the Middle Ages Christian priests who promoted the Crusades taught that it was virtuous to kill Muslims and in the modern world some Muslim religious leaders teach that it is virtuous to kill non-Muslims and even other Muslims.

The Spanish Inquisition tortured and killed people for the good of their souls and because the Church could claim half their victims' wealth. Many crusaders slaughtered Christians, Jews and Moslems indiscriminately and papal armies killed thousands of Cathar Christians in the Albigensian Crusade of the 13th century,[194] Christians in Europe and America tortured and burned women who were accused of witchcraft and in the modern United States some television evangelists collect offerings as representatives of God. Even now, some people demand that witches be burned and believe that anyone who does not obey the priests will spend eternity in hell.[195]

With the support and approval of imaginary gods, priests and other alphas can create the conventional wisdom of their choice. In ancient Greece, Aristotle taught that some men were natural-born slaves and others natural born masters.

In the modern world, many people claim to be atheists — according to one study nearly half of 728 students interviewed at Oxford University said they do not believe in any god[196] — but it can be argued that it takes more faith to be an atheist than to be religious.

It might someday be possible to prove the existence of a god or we can take the word of people who say they have seen proof, but while it might be difficult to prove that something exists, it is impossible to prove that something does not exist. Even on a local level , I challenge you to prove to me that there is no mutant crocodile in the sewers of Los Angeles. To do so you would have to conduct simultaneous

---

194 *Encyclopaedia Britannica CD 98*, "Albigenses"

195 http://www.landoverbaptist.org/sermons/witches.html See also https://www.landoverbaptist.net/showthread.php?p681875

196 Bullivant, Stephen, 'Where do atheists come from?' *New Scientist*, 6 Mar/2010, pg. 26-7.

inspection of every foot of every sewer in the city. Given a million or so volunteers, that might be possible — but how can you prove that something that you can't define or describe does not exist in the universe? I'm an agnostic because I don't have enough faith to be an atheist.

Back to the village. There may or may not have been a shaman in residence before the robbers moved in but, if so, he probably had prestige but little power. After the robbers move in a shaman can gain real power if he can legitimize The Baron because, in return, The Baron can give the shaman power and wealth.

The arrangement works even better if the robbers and their leader believe the shaman's claim to represent the gods, but the belief of the villagers is more important. Most rulers honor priests, but they don't have to believe in the religion. One section of Plato's *Republic* describes the god that must be worshiped by citizens of the ideal city state, without suggesting that such a god might actually exist.[197] Plato insists that mothers should be allowed to tell their children only approved stories, that it must be considered a serious sin to misrepresent the nature of the gods,[198] that the official god must not be seen as the author of all things, only of good things,[199] and that the approved religion must include a pleasant after-life, so soldiers would be willing to die on command.[200] When priests were not collaborating with rulers, they were competing with them and, at some times and places, the priests were more powerful and/or corrupt than the rulers.

Even now, some religious leaders promote hate. In July 2010, after Pastor Terry Jones of the ironically-named Christian Dove World Outreach Center in Gainesville, Florida, promised to burn 200 Qurans on the 2010 anniversary of the September 11 attacks, at least 20 people were killed in protests in the Middle East and Asia. Jones cancelled his threat and pledged never to burn a Quran but, on March 20, 2011, a "trial of the Quran" in his church found the scripture guilty of "crimes against humanity" and Pastor Wayne Sapp burned one in the church sanctuary. News of the desecration sparked protests in which at least 30 people — including seven United Nations workers — died. Jones insisted that he was not responsible.[201]

In Cleveland, Ohio, Samuel Mullet Sr., the leader of a renegade Amish sect, and 15 of his followers were convicted of orchestrating beard-and hair-cutting attacks that spread fear through the Amish of eastern Ohio. In five attacks, Mullet and his followers invaded private homes and, in the words of an AP news report, "sheared them almost like animals."

[197] Buchanon, Scot, ed, *The Portable Plato*, Viking Press Inc., NY, 1948, pg. 352-3.

[198] Ibid., p 353-4

[199] Ibid., pg. 357-8.

[200] Ibid., pg. 364.

[201] https://en.wikipedia.org/wiki/Dove_World_Outreach_Center_Quran-burning_controversy

Mullet is said to have forced errant followers to live in chicken coops and to have pressed married women to accept his intimate sexual 'counseling.'[202]

In England, members of a 'Muslim patrol' have been jailed for trying to enforce Sharia law on London streets. Wearing hoods, the 'patrol' forced men to dump alcoholic drinks, screamed at women who were not dressed according to Muslim custom and shouted insults at men they assumed to be gay.[203]

On March 11, 2002 15 young girls died in a school fire because officers and men of the Saudi Committee for the Promotion of Virtue and the Prevention of Vice would not allow firemen into the school, and would not allow the girls to leave because they were not wearing headscarves or accompanied by a male guardian.[204]

On March 19 of 2015 a 27-year-old woman scholar and teacher at a mosque in Kabul was beaten, stoned, deliberately run over by a car and finally set on fire because a Mullah said she had burned a Koran. In fact her crime was that she told women they did not have to buy the amulets the Mullah sold.

Thirteen policemen who stood by and watched the murder were suspended for not interfering.[205]

On Jan. 29, 2017 a 27-year-old university student took a rifle and a pistol into a mosque in Quebec City on a rampage that killed six people and wounded five others who were praying there. In previous incidents a pig's head was left on the doorstep of the same mosque, and another mosque was smeared with pigs' blood.

We assume that most low-ranking priests believe the myths themselves, but some may see them as tools to control the suckers who believe in them. It's possible that some high-ranking priests believe the myths too, but we might also suspect that plotting is more effective than piety as a route to the top.

Propaganda and Advertising

Some believers resent any suggestion that religious teaching and preaching are propaganda, but in fact the word is derived from the title of the *Sacra Congregatio de Propaganda Fide* (Sacred Congregation for the Propagation of the Faith) established by Pope Gregory XV in 1622 in an attempt to counter the Reformation. In 1982, Pope John Paul II changed the name to *Congregatio pro Gentium Evangelizatione* because the word has acquired such a pejorative connotation.

In fact propaganda is just persuasion, and in early human history it would have included one man's attempt to convince his cohorts that

[202] "Leader of Amish sect guilty of hate crimes," *Toronto Star*, Sept 21/12, pg. A21.

[203] http://guardianlv.com/2013/12/muslim-patrol-sent-to-jail-for-enforcing-sharia-law-in-london-streets/

[204] https://en.wikipedia.org/wiki/Committee_for_the_Promotion_of_Virtue_and_the_Prevention_of_Vice_(Saudi_Arabia)#Alleged_abuses

[205] "Murder in the streets of Kabul" Oates, Lauren, *National Post*, Mar 20/15, pg. A9.

they could eject that cave bear from their home — or that if they could kill a mammoth, they would have food for weeks.

Since the development of writing, propaganda has been put on paper in such well-known works as the gospels of the *New Testament*, *The Federalist Papers*, Payne's *Common Sense*, and Martin Luther's *Ninety-Five Theses*. Some sophisticated propagandists have buried their messages in fictional poems and stories, as Kipling's promotion of colonialism and Dickens' diatribes against the poverty and living conditions of the poor in his day. During World War II some of my favorite reading was a series of boys' books about American Dave Dawson and his English friend Freddy Farmer who, at one time or another, served in nearly every branch of the English and American forces. My book *Rescue Trooper* was written as deliberate propaganda to promote my idea that a modern nation should have an organized corps to cope with the effects of natural disasters.

French scholar Jacques Ellul distinguished two types of propaganda with two subdivisions of each. 'Agitative' propaganda tries to convince us to change something and 'integrative' propaganda tries to convince us that there is no need for change. Either type can be 'political' propaganda, which is disseminated by a political entity with a specific program, or 'sociological' propaganda, which is disseminated by all of us, often unintentionally.[206]

When a leader appears at a rally and his supporters cheer, that's both an example and a result of integrative sociological propaganda. The supporters cheer to show that they like their leader and, by cheering, they hope to encourage others to like him. When a union leader tells workers that the company makes too much profit and that they should demand higher wages, that's an example of agitative political propaganda. The distinctions between agitative and integrative and political and sociological propaganda are useful but, as we will see, they can be and often are blurred.

Sociological propaganda is generally benign but all kinds of propaganda are potentially dangerous because propaganda affects everybody, including the people who create it. The peacock with the most impressive tail probably believes he is the best mate and a good car salesman believes that the car he promotes is the best and that his customer needs it. A bible-thumping preacher may actually fear his own damnation and Maya priests and kings subjected themselves to painful rituals to please their gods.

We need to understand propaganda and to realize that it is not necessarily bad or evil or harmful. If I stage a rally to promote my idea that we should all be nice to one another, that's political propaganda, but it may not be harmful. If you go to that rally and cheer enthusiastically, that's sociological propaganda and, again, not necessarily evil.

[206] Ellul, Jacques, *Propaganda: The Formation of Men's Attitudes*, Alfred A Knopf, NY, 1965, trans by Konrad Kellen and Jean Lerner, pg. 109.

In the modern Western world, most propaganda is commercial. If I publish material that urges you to buy an Acme model XYZ gugglerump, or to buy any make or model of gugglerump from the Galumptious Gugglerumpery, that's advertising. If I publish an article about the wonderful device we call a gugglerump and how it can improve your life, that's propaganda.

Collateral Propaganda

Ellul recognized only two basic types of propaganda, but I recognize a third that I call *collateral propaganda*. Like sociological and political propaganda, it need not be harmful but I think it is more common, and often more harmful, than other kinds.

Like 'collateral damage' in a war, collateral propaganda spreads beyond its intended target and affects people it has not been aimed at.

Suppose, for example, the police chief of a town wants a new cruiser or more men or better radios or whatever. In order to convince people of his need he talks about drug dealers moving into town, or 'terrorism' or some other popular bogeyman.

Whether the town council listens or not, others will; and some will worry about this new threat. Someone may decide not to open a new business or build a new house because the character of the town is changing. A teenager may decide that if drugs are going to be available he will look for them, and a drug dealer from another town may decide that if things are happening in this town, he wants in on the action.

The prediction of 'terrorism' is particularly dangerous because many people who could become 'terrorists' have already given up control of their own lives and chosen to follow a leader rather than make their own decisions. Still, it's popular now because 'me too' politicians are eager to jump on the bandwagon.

The police chief may really need a new cruiser or a better radio but, in his campaign to get it, he may do real damage to the town.

The danger of propaganda is that it tends to stray further and further from the truth because the people who create it are, even more than the rest of us, continuously immersed in it. Because they live in a maelstrom of propaganda, their point of view is skewed and when they produce their own advertising or propaganda, they start from the skewed viewpoint.

It's tempting to think of advertisers and propagandists as Machiavellian geniuses who know exactly what they are doing and who intentionally lead us astray. Such people may exist, but most advertising and propaganda are produced by people like you and me who have a job to do and who are themselves fooled by the misinformation that assails us all.

Suppose that, as a propagandist, I aim to puff the truth by about 10 percent in everything I write for my client. When I write about Acme gugglerumps I puff your need for a gugglerump by about 10 percent, I puff the quality of the Acme brand by about 10 percent and I puff the benefits to the economy when you buy a gugglerump by about 10 percent.

If I were the only one doing it that would be no problem because I know the truth and when I write the next ad or public relations release, for gugglerumps or some other product, I will start from the same base.

But I am not alone. After I write my ad, I go out into the world where I am assaulted by literally thousands of other ads and promotional articles, all with a 10 percent puff factor. After a while I will accept that 10 percent puff factor as reality and, when I write my next ad or release I will puff the facts 10 percent from my new perceived reality, thus producing a puff factor of 21 percent. Because thousands of other ad-writers are doing the same I will soon accept this as reality and, from then on, when I puff my perceived truth by 10 percent my actual puff factor will be over 30 percent.

If we had a stable baseline to compare things with, we could see this happen, but there is no baseline. We live in a culture of puff and propaganda and, with positive feedback, it is running amok.

Barnum's Law

We see another type of collateral propaganda in the phenomenon I call *Barnum's Law*, in honor of American showman Phineas Taylor Barnum — proprietor of Barnum's museum in New York and a partner in the Ringling Brothers, Barnum and Bailey Circus, AKA "the greatest show on Earth."

Barnum was one of the great salesmen of history and he was famous for the way he cooperated with the press. Most businessmen will cooperate on a story that reflects credit on them but Barnum would help reporters on any story, good or bad. When people asked him why he would help reporters with a story that made him look bad, he stated his principle — which I consider one of the great laws of salesmanship — "There is no such thing as bad publicity."

He knew that whatever the story and whatever the public reaction to it, people would soon forget the story but they would remember the name.

Most businessmen will cooperate on a story that reflects credit on them but Barnum would help reporters on any story, good or bad.

In 1866, Henry Bergh, president of the American Society for the Prevention of Cruelty to Animals, wrote Barnum with a complaint that snakes in his New York museum were fed live rabbits. Somehow, Bergh's letter and Barnum's reply both found their way into the *New York World* newspaper and, as a controversy developed, other newspapers in the United States and even in England printed a public correspondence that continued from December of 1866 to June of

1867. Bergh's florid descriptions of the terror of the rabbits and the hypnotic gaze of the snakes were overdrawn, to say the least, but they were wonderful advertising and visitors flocked to the museum.

A friend of mine saw Barnum's Law in action when she worked for a driving school in Hamilton, Ont. When one of the school's instructors got drunk and drove one of the school's cars off Hamilton Mountain, the local press named the school in their stories.

Business dropped for a couple of weeks, but then it came back and it boomed for a couple of months.

An instructor who gets drunk and wrecks a car is not good publicity for a driving school, and as long as people remembered the accident, they stayed away from the school. But after they forgot about the accident they still remembered the name and, when they looked for a driving school, one name was more familiar than others.

Donald Trump reported his experience with Barnum's Law in *The Art of the Deal*. He had to tear down the old Tiffany Building to build the Trump Tower, and some people wanted him to save some sculptures on the building. Trump didn't think they were worth much and he knew it would cost him a small fortune to save them, so he had them destroyed.

The next day the *New York Times* ran a front-page picture of workmen destroying the sculptures and for several days the media and public figures criticized Trump for his decision.

"Even though the publicity was almost entirely negative," Trump wrote, "there was a great deal of it, and that drew an tremendous amount of attention to Trump Tower. Almost immediately we saw an upsurge in the sales of apartments."

He concluded that "good publicity is preferable to bad but, from a bottom-line perspective, bad publicity is better than no publicity at all."[207]

Thirty years after this experience Trump ran for president of the United States, against the opposition of much of his own party, with a social media campaign that insulted and alienated the established media. The media responded by giving him much more publicity than it gave his opponent, and Trump won the presidency. I'm not in the business of judging presidents but I think it's safe to say that Trump understands the media better than the media understands itself.

Barnum's Law often turns propaganda around, producing an effect opposite to what the propagandists expect.

The United States government got a lesson in this with the Eighteenth Amendment to the Constitution, which made it illegal to produce, transport or sell alcoholic beverages, and the separate Volstead Act which set down methods of enforcing the Eighteenth Amendment and defined which 'intoxicating liquors' were prohibited and which allowed for medical and religious purposes. The

[207] Trump, Donald, with Tony Schwartz,, *The Art of the Deal*, Random House, 1987, paperback by Warner Books, 1989. pp 174-176.

Amendment was ratified on January 16, 1919, to take effect on January 17, 1920.

One of the 'side effects' of Prohibition was alcohol poisoning. Since bootleg alcohol was not produced in distilleries under government supervision and not often made under the direction of chemists or qualified distillers, its quality was extremely suspect. The chances of obtaining 'real stuff' were never better than eight in a hundred! Like modern 'street' drugs bootleg booze was often spiked with chemicals and poisons to give it 'kick,' and deaths from alcoholic poisoning increased dramatically.

Real whiskey could be obtained by prescription from medical doctors. The labels clearly warned that it was for medicinal purposes and that any other use was illegal, but doctors freely wrote prescriptions and drug-stores filled them without question. No attempt was made to stop this practice, and many people got their booze this way.

More than million gallons of alcoholic drinks per year were sold by prescription and nobody knows how much was made illegally or smuggled into the country. Some of the smugglers became nationally-recognized entrepreneurs and even philanthropists in both Canada and the United States. A gang war in Chicago made Al Capone a national — even historic — figure and Prohibition is credited with the development of organized crime. It also it helped the FBI grow from a small and relatively innocuous office mostly concerned with commercial fraud to the large, aggressive and well-armed force it is today.

We see Barnum's law at work again in the American War on Drugs which has, in effect, publicized the illegal drug trade in the U.S. for nearly fifty years; giving the United States the world's worst drug problem and the world's biggest per-capita prison population; and created drug producing and smuggling cartels which have destabilized parts of neighboring Mexico.

If it were not tragic it would be amusing to compare drug use in the U.S.A., where tens of thousands of policemen and civil servants try to prevent the use of drugs, with drug usage in The Netherlands where marijuana is tolerated. The numbers are published in the pamphlet *Q&A Drugs, A Guide to Dutch Policy*, published by the Netherlands Ministry of Foreign Affairs and available through any Dutch consulate.

In 1997, 32.9 percent of Americans 12 years old and older but only 15.6 percent of Dutch citizens in the same age group had used cannabis at least once. Nine percent of Americans but only 4.5 percent of Dutch citizens had used cannabis in the year before the survey. More than 5 percent of Americans but only 2.5 percent of Dutch citizens had used cannabis within a month before the survey.

Lax drug laws bring thousands of American drug users to The Netherlands as tourists but, despite that, the country now has the lowest usage of hard drugs in the developed world.

Hard drugs are illegal in The Netherlands but, with no government campaign to publicize them, they are not a serious problem. More than 10 percent of Americans but just over 2 percent of Dutch citizens have tried cocaine at least once. Within the month before the survey 0.7 percent of Americans but only 0.2 percent of Dutch had used cocaine. Americans are more than three times as likely as Dutch citizens to have used heroin at least once.

The 'war on drugs' has given the United States the world's largest per-capita population of prisoners. With 4.6 percent of the world's population, the country that calls itself the 'land of the free' has 22.5 percent of the world's prisoners.

A website maintained by the United Kingdom's Home Office lists, among other things, the ratio of prisoners to population for most countries of the world. It says the United States holds 723 of every 100,000 of its population in jail, Canadian jails hold 116 of every 100,000 Canadians and the incarceration rates in Germany, Italy, France, Switzerland, Sweden, Denmark, Finland and Norway, (in that order) vary from 98 down to 59 prisoners per 100,000 population.

Meanwhile, the War on Drugs continues — with 'public education' campaigns that convey the message that drugs are readily available and that a lot of people find them attractive while police action supports the street price of illegal drugs and guarantees fat profits to dealers who can avoid arrest. The odds and the profits are much better than in any legal lottery.

In Canada we see Barnum's Law in the family violence industry, which was kicked off in 1980 by the report *Wife Battering in Canada* prepared for the Canadian Advisory Council on the Status of Women, and a follow up report on *Violence in the Family* presented to the Commons Standing Committee on Health, Welfare and Social Affairs in May of 1982. The cover of the report to the House of Commons featured a drawing of a man beating a woman and the introduction to the report contains the statement "We have been given good reason to believe that every year in Canada one tenth of the women who live with men as a couple are battered."

The report admitted that the 10 percent figure was "an estimate" but in fact there were hard numbers available at the time. As part of the Solicitor General of Canada's *Canadian Urban Victimization Survey* more than 61,000 Canadians in seven cities were interviewed and the survey results — published in 1983 but available earlier (and, I've been told, offered to the authors of the wife battering report[208]) showed that only 70 women per 1,000 of population, or less than 7 percent, were involved any form of violence in the year before the survey.

Because of the fuss caused by the 10 percent estimate the Solicitor General's next survey, published in 1985, went into more detail.

[208] I was told this by the mathematician who compiled the numbers for the Solicitor General's report and who personally offered his results to the compilers of the *Violence in the Family* report.

It found that 39 women per thousand, or less than 4 percent, were victims of assaults in the year, and that 35 of the 39 assaults were by persons other than spouses.

In the year of the survey four women per thousand, or 0.4 percent of the population, were assaulted by spousal partners.

The difference between the 0.4 percent found by the survey and the 10 percent estimated by the pressure groups is easier to understand if you know that the women were lobbying for a $20 million grant to open 'shelter houses' for battered women. It's much easier to get a grant for a facility needed by 10 percent of all women than for one that might be needed by 0.4 percent.

If it were just money, we might laugh it off because other groups have tapped the government for more than $20 million on false pretenses, but in this case the problem goes farther. Most Canadians laughed when the 10 percent figure was first quoted — even in the Commons some members laughed out loud — but the women used the famous technique of repetition. Through the media they hammered the figure home until it was finally accepted.

In obedience to Barnum's Law, media and government campaigns against imaginary cases of wife beating helped to create real cases. If it's unthinkable to hit my wife, I won't do it, but if 10 percent of men beat their wives, it can't be all that bad. Meanwhile the propaganda teaches women to expect to be beaten.

That's bad news for women but it's good news for the hate-mongers. By the early 1990s Canada's federal government was spending about $70 million a year on shelter services, counseling, police and education programs in support of 'abused women,' and was planning to increase the total to $136 million a year.

This is one area in which Canada can boast of world-class numbers. A study by University of Calgary sociologist Dr. Eugen Lupri and Elaine Grandin, also of the University of Calgary, found that the incidence of most violent crimes is about five times as high in the United States as in Canada but the incidence of family violence is much higher in Canada. Canadian men "kicked, bit or hit with a fist" their mates about ten times as often as American men, and Canadian women attacked their mates several times as often as American women. The only area of family violence in which Americans lead is in the use of guns.[209]

In 1992 then-Ontario Minister of Women's Rights Marion Boyd claimed that 'research' showed that one man in five, or about 20 percent, admit to using violence against the women they live with. Presumably the rest of us deny it.

Actually the number Boyd should have quoted was 12 percent, as revealed by Dr. Lupri in the Statistics Canada publication *Canadian Social Trends.* His study found that 12 percent of men interviewed

[209] Grandin, Elaine and Eugen Lupri, "Intimate Violence in Canada and the United States: a Cross-National Comparison," *Journal of Family Violence* Vol 12 #4.

admitted that they had pushed, grabbed or shoved their partners at least once in the preceding year.

The study was based on questionnaires distributed to 471 men and 652 women, but Statistics Canada chose to publish only the numbers on male violence. For the record, here are some of the numbers contained in the full report.

- 9.1 percent of husbands had threatened to hit or to throw something at their mates in the previous year, and 15.9 percent of wives had done the same.

- 11.9 percent of husbands and 13.1 percent of wives had pushed, grabbed or shoved their mates, 5 percent of husbands and 7.6 percent of wives had slapped, 5.4 percent of husbands and 7.6 percent of wives had hit or tried to hit, 6.4 percent of husbands and 6.3 percent of women had kicked, bit or hit with a fist, 2.5 percent of men and 6.2 percent of women had physically beaten up a partner

- 2.1 percent of men and 3.6 percent of women had threatened their partner with a knife or a gun. One half of one percent of men and 0.8 percent of women had actually used a weapon.[210]

As might be expected, home violence has increased to keep pace with the bigger grants and hate-mongers still indulge in poetic license when they quote or interpret the numbers. When a Statistics Canada paper issued in the summer of 1997 reported a decline in wife-beating, women's groups said that meant women were no longer reporting assaults.

Many of the people who run propaganda campaigns know enough about human nature that they should understand what they are doing but they may be blinded by hatred, self-interest and their own propaganda.

In 1987 the propagandists found a new opportunity in the sexual harassment business. That year the British Columbia Public Interest Research Group, a student organization at Simon Fraser University, published a *Report on Sexual Harassment and Sexual Assault at Simon Fraser University*.

The original plan was to survey all female students at the university and the first questionnaire was published in a student newspaper called *The Peak* but the newspaper survey produced, in the words of report author Anne Burger, "negligible" returns.

[210] Lupri, Eugen, "Hidden in the home, the dialectics of conjugal violence," (English version of "Harmonie und Aggression: Uber die Dialektik ehlicher Gewalt," *Kolner Zeitshcrift fur Soziologie und Sozialpsychologie*, 42 (3) 474-501, 1990 see also Grandin, Elaine and Eugen Lupri "Intimate Violence in Canada and the United States, a cross-national comparison" *Journal of Family Violence*, Vol 12 #4, 1997; and Grandin, Elaine, Eugen Lupri and Merlin B Brinkerhoff, "Couple Violence and Psychological Distress," *Canadian Journal of Public Health*, 89, Number 1, (Jan/Feb, 1998, pp 43-47.

Other researchers might have taken this as a comment on the women's perception of sexual harassment but Burger and her crew were not to be daunted. As an alternate strategy they gave the questionnaire to a selected group of 350 undergraduate women, watched them complete it, and then took it back. Working this way they got 346 completed questionnaires for a response which their report interprets as 98 percent.

The report does not say how the participants in the group were selected, but it is reasonable to assume that they were not chosen for either their opposition to radical feminism or their immunity to propaganda.

Burger also placed questionnaires and letters explaining the project in the mail boxes of 444 graduate students, and got back 98 of them for a response she describes as 20 percent.

The total sample was 794 of a total of about 4,700 women students in the university. The report says the respondents were randomly selected but that's obviously not true. The sample included more graduate students than undergraduates and almost all the undergraduates were freshmen who filled out the questionnaire under supervision. Undergraduate second, third and fourth year students, who make up the majority of all students in the university, were not represented.

Within the selected group a total of 237 respondents were judged to have been sexually harassed. The research group interprets this as 53 percent, because they counted only the questionnaires that were returned. If we interpret the 80 percent of questionnaires not returned by graduate students as a 'no' answer, the percentage who have suffered harassment drops considerably.

But even a lower number may be open to question because when we look at the questions we find that a women has been 'harassed' if she has ever heard a discriminatory remark, including a joke, in a classroom. The way the question was worded the joke or the remark did not have to be directed at a woman. If a woman overheard one man tell another an off-color joke, or comment about some feminine characteristic, the woman was 'harassed.'

Several women were 'harassed' because they were afraid to move around the campus at night and one was 'harassed' because she felt she had to be "un-naturally unfriendly" to men. She did not say whether it was men or women who convinced her that she had to be unfriendly to men.

Most of the students who were 'harassed' did not seem to consider the incidents serious. Of the 235 who said they had been harassed 144 — more than half — answered "does not apply" to the question that asked if they reported it. The report says it assumes the students did not consider the incidents to be important. Every incident had to be important to Burger, of course, because without them she would have had no report.

Other researchers appear to be willing to tolerate sexual harassment, provided it is men who are harassed. In 1987 — the same year as the BCPirg report — doctoral candidate Gina Fisher studied sexual harassment at the Ontario Institute for Studies in Education.

She mailed questionnaires to all the 527 women and 235 men students at OISE and got responses from 239 women, 83 men and 30 people who did not identify their sex. The ones who did not specify their sex were not considered in sex-difference analyses.

The returns indicated that 56 percent of male respondents thought they had been harassed, compared with only 51 percent of women. More men than women reported harassment in the sense of being pressured for dates, but Fisher chose to explain the figures away.

"It was speculated," she says in her report, "that the men in this study may have misinterpreted nonsexual behaviors by female students as sexual in nature. Hence males' tendency to oversexualize women's friendly overtures may have led male students to report high rates of sexual harassment."

After discounting the men's complaints, Fisher found "no sex difference in reported sexual harassment incidence rates," thus demonstrating one researcher's tendency to make the facts fit her theory.

But Fisher did find that "a significantly greater number of women than men defined 15 of the 20 unsolicited sexually-oriented kinds of behavior presented to them as sexual harassment.

One difference was that men were likely to consider even an unwanted proposition as flattering, but most of the women considered it an insult. The report says that 38 percent of women rated personal invitations — like a request for a date — as offensive and interfering.

Fisher found that women were less tolerant of harassment by professors than were men and she admitted that some women were offended by "even behaviors that may seem relatively benign."

Men are more tolerant than women, in other words, but even so they complain of more harassment. A researcher other than Gina Fisher might have concluded that men suffer more harassment.

The fact is that years of propaganda and mock science, like Burger's and Fisher's reports, have made a lot of women fear men and, for these women, the existence of men is seen as harassment.

Despite the evidence of her own study Fisher's report, which started with the premise that women need to be protected from harassment by men, ended with the conclusion that women need to be protected from harassment by men.

And women are protected, as Professor Richard Hummel of the University of Toronto discovered. In the fall of 1988, a female part-time student complained that then-60-year-old Hummel "stared at her" while they both swam in a pool at the University of Toronto.

The university's official policy on sexual harassment requires mediation of such complaints but this student demanded a formal hearing with no attempt at mediation, and Sexual Harassment Officer Nancy Adamson complied. It has been suggested that the reason she skipped a few steps of the required procedure was that her first annual report was nearly due, and she had no other activity to report.

Whatever the reason for the hearing Hummel was convicted, barred from the university's recreation complex for five years, and ordered to counseling approved by the sexual harassment officer.

The case became a laughing stock because the student had charged that Hummel stared at her while they were both swimming, which is obviously impossible, but the effects were not funny.

Hummel's name had been smeared and he was the victim of a hate campaign which included vicious graffiti (one said "poke his eyes out") and someone stuck pins in the tires of his bicycle. He tried to take his case to a real court but when the court date approached Adamson took a six-month vacation in Tahiti and was not available to testify.

As Hummel fought to regain his reputation the University of Toronto administration fought back. After several years of bitter feuding, Hummel accepted early retirement. The case cost the university hundreds of thousands of dollars — perhaps millions — and the services of a senior professor.

And the Hummel case is just one of many. The Ontario Human Rights Commission alone held 139 hearings in 1990, and Ontario is only about a third of Canada — which in turn is about one tenth of North America. When you consider cases like Hummel's — which was tried in secret by a panel at the university and which does not show in the official figures — the total number of hearings and trials in North America may be in the thousands.

But the prohibition against harassment works only one way. At Toronto's Ryerson University, where posters advise women to complain about any comments they don't like, the campus bookstore once featured *The Dumb Men Joke Book* prominently displayed beside the cash registers. If the sex were changed, any one joke in that book would be cause for complaint by women.

Maclean's Magazine reported the case of Carleton University first-year psychology student Lyle Burwell, who complained about offensive cartoons in the student newspaper. University associate vice president Marilyn Marshall, who handled the complaint, explained that the cartoons — including one in which a smiling woman with a knife asks women whether their lives would be helped by the "total elimination of penises" just illustrate female fantasies and they are "not the same as endorsing or promoting the fantasized action."

One might wonder whether she would be equally philosophical about a cartoon of a naked woman tied to a whipping post.

The sexual harassment business provides hundreds of well-paid jobs for sexual harassment officers and staff at nearly every university and community college, millions of dollars' worth of business a year for lawyers, and something for the media to report but, like many modern businesses, it produces nothing of value to society. In fact it probably does serious harm.

Common sense suggests that it produces more sexual harassment, or sexual harassment where there would otherwise be none. Let's not forget that most students ignored Burger's first questionnaire, published in the student newspaper, and the only way she got the results she wanted was to have selected students fill out questionnaires under supervision.

That does not mean there was no behavior in those days that would be considered sexual harassment today, but it implies that the problem was not serious. When comments were made, women were not sensitized to react. Perhaps they laughed them off, the way Carleton's assistant vice president expects men to laugh off cartoons that depict her fantasies of mutilating men.

But women don't laugh off insults now, because the hate propaganda sensitizes them, and they probably get more insults and other forms of harassment than ever before because, consciously or not, men resent the hate propaganda,

"After an experience of sexual harassment/assault with a particular man," Burger wrote in her report, "respondents invariably adopted negative feelings toward men in general." Burger and other militant feminists seem to ignore the possibility that after being nagged for years on end by professional man-haters, some men may be less than sympathetic to women's cause.

And some may go to extremes. On Dec. 6 of 1989, Marc Lepine took a semi-automatic rifle to the École Polytechnique engineering school in Montreal, where he shot and killed 14 women. Lepine avoided shooting men, and he screamed his hatred of women as he shot them.

For feminist propagandists, of course, this is proof that they were right — not that they were wrong. Granted that Lepine was not of sound mind when he did the shooting, we must also question the sanity of the supposedly-qualified psychologists who mounted the hate campaign that set him off.

We see another example of Barnum's Law in the Canadian Firearms Registry, which was supposed to make Canadian streets safer by making it more difficult for criminals to buy guns but which has, instead, sparked a frightening increase in the use of guns in crime in Canada. Established in 1995 and partly repealed in 2012, the registry changed the nature of crime, and of city life, across the country.

The explanation is simple. The gun registry made guns a prestige item for suggestible people on the edge or the wrong side of the law. With guns readily available in the United States and a notoriously porous border, a black market quickly made guns available on virtually every street corner of large cities.

When guns began to appear on Toronto streets they were news, and both the media and supporters of the registry reacted with glee. To the media this was worthy of front-page coverage and to promoters of the gun registry it was proof of what they had been telling us — that guns were an essential accessory to crime. The message was obvious — that any would-be criminal or tough guy who didn't have a gun was, somehow, inadequate.

In 2009, Toronto City Police Staff Inspector Kathryn Martin told a Globe and Mail reporter, illegal handguns were a rarity in Toronto in 1985.

"If somebody seized a gun," she said, "the whole platoon would come in off the road and look at it."[211] Now pistols and even sub-machine guns are standard equipment for teen-age gangs and shootings are routine events.

When the registry was established in 1995 proponents said it would cost a total of $119 million, but that fees would cover most of the cost and taxpayers would pay only about $2 million. Final estimates of the cost, after it was abolished, set the price at somewhere between $1 and $2 billion.

Modern salesmen may not admit that they believe in Barnum's Law, but we see it in action every day. Think of all the TV commercials that are famous because they are 'dumb,' and because people resent them. The people who run those commercials measure their results, and they know that most people will remember the name long after they forget the commercial.

TV commercials are powerful motivators but, because they are powerful and the people who produce them don't worry about side-effects, they can do a lot of harm.

Each individual car dealer, for example, advertises the cars he sells but, together, they all promote the idea that everyone needs a new, big and expensive car. They know that not everyone needs or can afford one, of course, but they don't know or care that their combined propaganda, plus the national and regional advertising and propaganda of the companies they represent, creates needs that can't be filled and therefore must make some people unhappy.

The perceived need for a new car may not be a serious problem, but remember that virtually every business is trying to establish a need for its wares and while any single unfulfilled need of this sort may be a small matter, the combination of all of them, and the social isolation felt by those whose perceived needs are unfilled, is not. Most poor people in most first-world towns and cities have more material goods than most of the hunter/gatherers in the world, but they are taught by their society that they need more material goods and that they can't be happy — or accepted by the people they see as their peers — without them.

We all create collateral propaganda, to a certain extent. The preacher and the manager of a charity who talk about the misery of

[211] http://www.theglobeandmail.com/news/toronto/guns-the-new-life-on-the-streets/article1201916/

the poor, and all of us when we talk about — or perhaps hint at — our personal status and prestige, are forming or trying to influence others' view of the world.

Propaganda and Education

Some people think their education and study of 'news' keep them safe from propaganda, but that's not so. A hermit who lives in a cave on a remote mountain might be immune to modern propaganda but the rest of us are all affected one way or another. As a general rule, the more we are part of society the more propaganda we are subjected to and the more likely we are to be influenced by it. As Jacques Ellul wrote ...

> The uncultured man cannot be reached by propaganda. Experience and research done by the Germans between 1933 and 1938 showed that in remote areas, where people hardly knew how to read, propaganda had no effect. The same holds true for the enormous effort in the Communist world to teach people how to read. In Korea the local script was terribly difficult and complicated: so, in North Korea, the Communists created an entirely new alphabet and a simple script in order to teach all people how to read. In China, Mao simplified the script in his battle with illiteracy and in some places in China new alphabets are being created.[212]

And, a page later in the same book (italics Ellul's)

> The number of propaganda campaigns in the West which have first taken hold in *cultured* settings is remarkable. This is not only true for doctrinaire propaganda, which is based on exact facts and acts on the level of the most highly developed people who have a sense of values and know a great deal about political realities, such as, for example, the propaganda on the injustices of capitalism, on economic crises, or on colonialism; it is only normal that the most educated people (intellectuals) are the first to be reached by such propaganda. But this is also true for the crudest kind of propaganda; for example, the campaign on Peace and the campaign on bacteriological warfare were first successful in educated milieus. In France, the intellectuals went along most readily with the bacteriological warfare propaganda. All this runs counter to pat notions that only the public swallows propaganda. Naturally, the educated man does not *believe* in propaganda; he shrugs and is convinced that propaganda has no effect on him. This is in fact one of his great weaknesses and propagandists are well aware that in order to reach someone, one must first convince him that propaganda is ineffectual and not very clever. Because he is convinced of his own

[212] Ellul, Jacques, *Propaganda, the formation of men's attitudes*, Alfred A Knopf, NY, 1965, trans by Konrad Kellen and Jean Lerner, pg. 109-110.

superiority, the intellectual is much more vulnerable than anybody else to this maneuver, even though basically a high intelligence, a broad culture and a constant exercise of the critical faculties and full and objective information are still the best weapons against propaganda.[213]

Advertising

Advertising is a close relative of propaganda and it too has been around for a long time. In the natural world flowers advertise their nectar and pollen with bright colors and sweet smells and male birds advertise their eligibility for mating with bright colored feathers, songs or other means.

A Babylonian clay tablet that has been dated to 3000 BC advertises the services of a shoemaker, a scribe and an ointment dealer. Some shops in Pompeii had inscriptions on walls near the entrance to inform passers-by about the products they offered and ancient Egyptian traders used papyrus for wall posters. Commercial messages were and still are painted on walls rocks in many parts of Asia, Africa, and South America.[214]

The best modern advertising bypasses rational thought processes by presenting images, rhythms and managed emotions rather than ideas.

In one famous advertisement, a handsome and obviously healthy cowboy rode a beautiful horse through beautiful countryside. The scene had nothing to do with the cigarettes it advertised but it portrayed a life that many people idealized, and the one part of it that anyone could achieve was to smoke the advertised cigarettes. Millions of people reacted to that picture and accepted the cigarettes as part of the scene.

Even in nature, some advertising is false. Brilliant red, yellow and black colors warn would-be predators that the coral snake is deadly, and some harmless snakes show the same colors in almost the same pattern. Several varieties of syrphid flies scare away predators by mimicking the color and behavior of bees. Monarch butterflies are safe from birds because they have a foul taste and the Viscount, which is edible, is safe because it looks like a Monarch.

We normally see advertising as distinct from propaganda but there is a middle road, in the pseudo-editorial material we call 'advertorials.' As a general rule the makers and sellers of goods — gugglerumps, for example — create and pay for advertisements and the media in which they publish their ads creates and publishes advertorials, and these are separate operations. Granted that the people who make and sell gugglerumps may help to prepare the advertorial or even write it themselves, an advertorial will promote gugglerumps in general rather

213 Ibid., pg. 111.

214 http://www.yourarticlelibrary.com/advertising/essay-on-world-history-of-advertising-2382-words/22249/

than any specific make of gugglerump and it will at least pretend to present an unbiased view. Because the media depends on or hopes for advertising from several makers of gugglerumps the advertorials may include information or other material supplied by several or all of them.

Most advertorials are more or less accurate and truthful. Some people think of propaganda as 'a pack of lies' but a lie may be caught and, if it is, the propaganda will be nullified. The art of the sophisticated propagandist is to tell the truth, but only that part of the truth that he or she wants known.[215] As the poet William Blake wrote in *Proverbs* (line 95): "A truth that's told with bad intent beats all the lies you can invent." Still, some material that can be classified as propaganda might well be true.

Subliminal Suggestion

Advertising and propaganda make the modern world possible but they have a dark side, in the deliberate use of subliminal suggestion. It's not a new idea — the Greek philosopher Democritus recognized it about 400 BC, and Plato and Archimedes both recognized it.[216]

Thoreau wrote: "Many an object is not seen, though it falls within our range of visual ray, because it does not come within the range of our intellectual ray, i.e., we are not looking for it. So, in the largest sense, we find only the world we look for."[217]

Most Americans heard about subliminal advertising in the 1950s when a movie theater in New Jersey flashed messages, urging people to buy popcorn and Coca Cola, on the screen during a movie. The flashes were so brief that people did not consciously see or read them but, among nearly 50,000 people who saw them, sales of popcorn increased 57.7 percent over normal, and sales of Coke increased by 18.1 percent.[218]

After the test was publicized several American states proposed laws that would ban all subliminal advertising, but none were passed. When journalist Vance Packard recognized subliminal images in magazine advertisements he wrote a book about them. *Hidden Persuaders*[219] was a best seller, but the images are still there.

Writing in the early 1970s, Professor Wilson Bryan Key said a survey found that 60 percent of Americans believed that subliminal

[215] By this standard, of course, almost any newspaper article could be classified as propaganda. As we will see, that would not be completely unfair.

[216] Dixon, N.F., *Subliminal Perception, the nature of a controversy*, McGraw-Hill, London, 1971, pg. 6.

[217] http://www.walden.org/Library/Quotations/Observation. The website says he wrote this in his journal on 2 July 1857.

[218] Key, Wilson Bryan, *Subliminal Seduction*, paperback, New American Library, New York, 1974, pg. 22.

[219] Longmans, Green and Co, London, 1957.

advertising techniques exist only in science fiction and that 90 percent believe that they are illegal.[220]

That should be no surprise. If you were amoral and in a position to use such a powerful advertising tool, wouldn't you use it to convince people that it is illegal and/or that it doesn't work? I have no evidence that advertisers deliberately set out to convince people that they have nothing to fear from subliminal advertising, but the fact is that most of us believe we have nothing to fear while in fact we have a great deal to fear.

In 1971, English psychologist N.F. Dixon wrote that more papers had been published on subliminal perception than on any other subject in psychology, but most textbooks either ignore or minimize it.[221]

Any doubts that subliminal advertising exists and is used were laid to rest in the summer of 2000 when news programs around the world revealed and televised the subliminal content of one of presidential-candidate George W. Bush's TV commercials. When the commercial was slowed down the word "rats" was seen to drift across an image of Bush's opponent, Al Gore. Bush said he did not know about the subliminal message and some people may have believed him, but there is no question that it was there.[222]

Until that subliminal was discovered Bush led the polls by a wide margin but after it was publicized he lost so much popular support that when he was declared winner of the election many people questioned the win. I can't help wondering whether Bush slipped in the polls because he was caught using a subliminal or because he stopped using it.

In that case, the subliminal content was easy to find because it used an obsolete technique. Back in the 1970s, Key wrote that 'modern' subliminal techniques use constant low-intensity messages rather than a high speed flash, and they are much harder to detect.[223]

In fact it would be impossible to ban subliminals, because there are so many techniques and many of them are very subtle. We might also question whether a total ban on subliminals would be in the public interest.

In some American jails, for example, the walls of one cell are painted a specific shade of pink because men exposed to that color are less likely to fight.[224] Should it be illegal to use that color? Some chain

---

220 Key, Wilson Bryan, *Subliminal Seduction*, New American Library, New York, 1974.

221 N.F. Dixon, *Subliminal Perception, The Nature of the Controversy*, McGraw Hill, London, 1971), pp 34.

222 "Rats in ad dog Bush," *The Globe and Mail*, Sept 13/00, Metro edition, pg. 3. See also http:// www.youtube.com/watch?v2NPKxhfFQMs

223 Key, Wilson Bryan, pg. 23; see also Turnbull, Andy, *We Need to Talk*, Red Ear Publishing, Toronto, pg. 194 (2005 ed.), pg. 182 (2007). The book is available as a free pdf download at http://andyturnbull.com/ntt.pdf

224 Alter, Adam, *Drunk Tank Pink*, Penguin Press, NY, 2012.

stores bury subliminal messages — "I am honest, I will not steal" — in their background music, to reduce shoplifting.[225] The technique seems to reduce shoplifting — should it be banned? Maybe not — but what other messages are buried in the music?

Some colors make people aggressive, others make us feel cooperative. Some make us tense, some relax us. Should color be banned? Stores are laid out to encourage us to buy. Should store layouts be banned?

We can't control subliminal advertising, but we must learn to recognize it.[226]

The Subliminal Content of Entertainment

We tend to associate subliminal messages with advertising but there are at least three levels of subliminal messages in modern entertainment. They might be termed sociological rather than political propaganda, and I don't suggest that they are deliberate; but whether deliberate or not, they exist and they are not beneficial to society.

The most obvious is the choice of subject. Most people find violence and the threat of violence — crime or war — exciting, and it's no surprise that both are the topics of much of our fictional writing, movies and television. This is the choice of the producers and it's valid, in commercial terms, but it normalizes both crime and war — which makes both of them more acceptable and, we have to assume, more prevalent than they would be otherwise. It's also worth noting the popularity of modern video games, many of which teach young boys the pleasures of crime and war.

In this context it's worth reviewing the five factors — *primacy, authority, imitation, comfort* and *repetition* — that make us believe things, and noting that three of the five — *authority, imitation,* and *repetition* — are inherent in TV serials and popular movies and a fourth, *comfort,* clicks in when we believe what our friends believe.

Movies about crime normalize crime and violence, and some glamorize it. Indeed, most of our media normalizes and/or glamorizes crime and violence, and the effect is very powerful. As psychologists and propagandists know, theater and especially movies are the most effective means of delivering propaganda.

In 1895 French social psychologist Gustav Le Bon wrote, "The imagination of crowds is most stirred when ideas are transmitted dramatically. For this reason, theatrical representations . . . always have an enormous influence on crowds."[227]

We may not recognize the propaganda effect of fictional movies and TV but, in fact, fiction is a wonderful medium for propagandists

---

[225] Taylor, Eldon, *Subliminal Communication: emperor's clothes or panacea?*, Just Another Reality, Las Vegas, 1990, pg. 5, 35-7.

[226] http://www.gatheringspot.net/news-article/general-discussion/sound-silence

[227] Le Bon, Gustav, *The Crowd*, London, 1896, (1953 printing) p. 68.

because it can distort facts as required and if the distortions are noticed they are dismissed as 'artistic license.'

Fiction can use rhythm, emotion and other techniques to insert messages directly into our subconscious minds and, if there is any danger that we will think critically about anything, fiction can distract us with action or a love scene or a description sprinkled with emotionally loaded images or adjectives.

When fiction presents an argument it also presents the counter-argument, but it presents only the counter-argument that the author chooses. The result is that we see the point argued and won without hearing any real argument. Even the author of fiction may be fooled, because we can all convince ourselves of the truth of anything we really want to believe.

Much fiction actually depends on the 'willing suspension of disbelief.' Most of us know that a Swiss doctor named Frankenstein did not really make an artificial man, that Captain Kirk does not command a real starship and that Captain Nemo did not build and command a nuclear submarine in the 19th century but, for the sake of the story, we are willing to pretend belief. In all these cases the pretense is obvious and not likely to be mistaken, but when the story is more plausible we may forget that the 'facts' of the story are created in the writer's imagination.

This does not suggest that novelists and film writers are part of a deliberate plot to manipulate us. There is no need for such a plot because, in most cases, when an American writer presents the 'American Way of Life' as the best in the world, he actually believes that to be so. Most of the propaganda we find in fiction is integrative and sociological and the people who present it may be trying to lead but, in most cases, they are not trying to mislead their audience.[228]

It's hard to say whether the big change in media content was driven by world events or by new technology but, soon after World War II, the American entertainment media was flooded with war books and movies. The war was the most exciting thing that had happened to most Americans and it was an obvious topic for both writers and audiences.

But the flood of war stories normalized war and the United States that tried to avoid involvement in two world wars now seems to see war as a national sport. Small boys play at mock wars, and politicians drive the country into war after war. This is partly driven by the MIC, of course, but we can also say that the MIC is validated by the trend.

One popular theme of fiction glorifies a rogue police officer who ignores rules and shoots first. In the movies he's always right — the

[228] In fact the phrase "the American way of life" was an advertising slogan, used by the National Association of Manufacturers in a 1937-8 campaign to encourage Americans to spend more money despite the Depression. See Ewen, Stuart, *PR!, A social history of spin*. Basic Books, HarperCollins Publishers, NY, 1996, p. 320.

target is a bad guy who is about to try something on the cop — but, in real life the victim may be innocent.

In Marksville, Louisiana three city marshals and one policeman stopped a civilian's car, shot and wounded the unarmed driver and killed his six-year-old son.[229]

If it had been a movie, the driver would have turned out to be a terrorist and the son an alien invader from outer space.

In fact there seems to be some question whether the marshals had any right to stop the car at all. Apparently their legal function is as process servers, but they took up traffic enforcement about three months before the fatal shooting because the city cut back on the court's funding, and the marshals took up free-lance law enforcement to raise money to pay their wages. According to some accounts, the driver had his hands up when the marshals started shooting.[230]

A policeman in Cleveland shot a 12-year-old boy who brought his pellet gun to a public park. The gun looked real — but the policeman opened fire almost immediately, without giving the boy a chance to drop his toy.[231]

It's hard to blame the policeman because he has probably seen movies or TV shows in which a policeman who was slow to act was shot and/or the policeman who shot first was a hero. We might wonder about the parents, and the society, that thinks a toy gun is an appropriate plaything,

False Memories

The movies and TV shows were fictional but it's quite possible that the policeman remembered them as real events. We know the fiction we read in books is fiction because we have to use a learned skill — reading — to extract ideas and visualizations from a printed page, but it's easy to confuse modern high definition color movies and TV with reality because we perceive and interpret them directly, using the same mental processes we use to perceive and interpret reality. We have even more potential for confusion because we know that some of the TV images we see are actually real-time broadcasts of real-life events.

---

229 http://www.npr.org/sections/thetwo-way/2015/11/07/455099803/two-officers-arrested-in-lousiana-killing-of-6-year-old-boy

230 http://www.nydailynews.com/news/national/la-state-police-probe-shooting-killed-son-suspect-article-1.2424300; http://www.scribd.com/doc/288823061/Statement-from-Marksville-City-and-Ward-2-Marshal-Floyd-Voinche-Sr http://www.theadvertiser.com/story/news/local/louisiana/2015/11/06/names-marksville-city-marshals-fatal-shooting-released/75318692/; http://www.cbsnews.com/news/jeremy-mardis-shooting-death-motorist-had-hands-up-as-marshals-killed-his-boy/; http://www.dailymail.co.uk/news/article-3317549/Father-autistic-boy-jeremy-Mardis-gunned-Marksville-Louisiana-police-DID-hands-opened-fire-reveals-eye-witness-backing-claim-bodycam-film.html

231 https://en.wikipedia.org/wiki/Shooting_of_Tamir_Rice

One of the most common failings of memory is what psychologists call 'source amnesia,' in which we remember something but are not quite sure where or how we learned it. Once the source of a memory is forgotten, we find it hard to distinguish between our memory of an event that we have actually perceived and of one we have only heard about.[232]

Psychology professor Elizabeth Loftus tells us that in several speeches former U.S. president Ronald Reagan cited a specific act of heroism in which a bomber pilot won the Congressional Medal of Honor. The incident he referred to is similar to one portrayed in a movie (with a navy pilot as the hero) but it does not appear on the records of any winners of the Congressional Medal of Honor.[233]

Sometimes it's hard to distinguish what we have learned from the news from what we have seen in real life. Several weeks after a sniper killed a child and an adult and wounded thirteen children and an attendant on the playground of an elementary school in Los Angeles, researchers interviewed 113 children to study their memories of the event.

One girl 'remembered' walking out to the playground with the girl who was later killed, and later she saw the dead girl, with the sniper standing over her body. Her memory was clear — but she could not have seen the dead girl from where she said she was standing, and the sniper never entered the playground.

World famous psychologist Jean Piaget had a clear memory of the time his nurse fought off a man who tried to kidnap him. He was five years old at the time, but when he was 15 the nurse admitted that the story had been a fabrication, made up to impress Piaget's parents. Piaget remembered the nurse telling his parents about the supposed incident, and for years he thought he remembered a real event.[234]

I have a distinct recollection of two friends telling me about psychologist Phillip Zimbardo's famous Stanford Prison Experiment in which, to study the inter-actions between prisoners and guards, he set up a make-believe prison at Stanford University. My memory of the discussion is clear, but the fact is that Zimbardo's experiment was conducted in 1971 and I haven't seen these particular friends since we worked together on an AFSC project in Mexico in 1956. My two friends were both recent graduates of ivy league universities and the source of many interesting conversations, but not of this one.

Obviously, our memories can play tricks on us. Quoted in the *San Francisco Chronicle*, psychologist Daniel Schachter of Harvard University says all our memories, whether we remember the source

---

232 "Amnesia to Confabulation — New Research on Memory," Daniel Goleman, *The San Francisco Chronicle*, 05/31/1994, final, pg. A7.

233 Loftus, Elizabeth and Katherine Ketcham, *The Myth of Repressed Memory*, St. Martin's Press, New York, 1994, pg. 93.

234 Ibid., pg. 97-99.

or not, are subject to what he calls 'leakage.' When we recall a scene or an event we may include details from a different scene or event.

The second level of subliminal content in entertainment media is in the prejudices, conscious and unconscious, assumed and promoted by a book, movie or television show. To some extent the characters in the narrative will reflect the view of the overall society — no surprise there, because the narrative is produced by members of the overall society and it is tailored to appeal to members of the overall society — but if the characters are impressive, some members of society will choose them as role models.

Is this a problem? I think it is, because some cultural stereotypes should be allowed to die a natural death. In the American South in the 1920s and 30s or in South Africa under apartheid a movie or story that portrayed a black man as intelligent or courageous would not have been a commercial success. The lead characters of a story have to reflect the ideals of the audience and, if they are impressive, some members of the audience will accept those lead characters as role models.

Police around the world see copycat crimes. Some are copied from movie or TV plots, and some movie or TV plots are copied from real life. I'm not promoting censorship here, but I think self-censorship is a good idea. I don't know that any real crimes are copies of fictional crimes, but I would be surprised if there were none.

It's also unfortunate that the heroes of modern fiction are mostly lawyers or policemen or secret agents — cultural heirs of the bandits that conquered and occupied the craft village and who live at the expense of the makers who actually create the wealth of the nation. This is understandable because most makers in the modern world are production-line workers, rather than craftsmen and, on average, takers are paid more than makers.

I don't question the need for policemen, lawyers, entertainers, businessmen and so forth in the modern world but they are still a cost to society and, for any society that has too many of them, a burden. Still, they have the most attractive jobs and life-styles and the fact that they tend to be glorified in fiction makes their work look even more attractive — so people are attracted to life as takers rather than as makers.

Modern media often justifies and glamorizes revenge and even sadism, but never the people who contribute to the real wealth of society.

The third level of subliminal advertising is product placement in movies and TV shows, an industry that is now approaching $2 billion a year in total sales. Product placement can work two ways — either the product I promote looks very good in the show or perhaps it easily out-performs a competitor. Either way, the advertiser provides the product free to the producers and may pay a considerable sum of money to have it portrayed as he wants it portrayed.

This is good business for the advertisers but questionable value to society as a whole. In movies and TV shows some minimum wage workers drive expensive cars and live in obviously expensive apartments, setting a standard that minimum wage workers in real life may expect to reach but cannot. High expectations must be good for advertisers, but they can be frustrating for people who can't meet them.

I have no evidence that a fourth level of subliminal content exists in modern media, but I suspect it may and I am sure that if it does not now, it will someday.

I have already mentioned the subliminal 'rats' message discovered in one of George W. Bush's campaign advertisements. It was discovered because it used a clumsy and outmoded technique, but better techniques are available.

Perhaps the best is the constant low-intensity write-over of an image or a sound. Using this technique commercial or political messages could be added to movies, TV shows, news and perhaps even concerts, effectively beaming opinions, ideas or prejudices into the audience's subconscious.

We like to think that we are ruled by governments that we control but there are also warlords — economic if not military — who ignore or bypass laws. Some of them have billions of dollars to spend and some already control media outlets. If they are not using subliminal messages in entertainment media yet, we can expect they will soon.

The Mind Benders

Most individual advertisers and propagandist are just trying to promote their own personal interests but they, along with religious leaders and some rulers, are members of the metasystem that I call the Mind Benders which is, perhaps, the real ruler of our culture. I do not suggest that the individual members of this metasystem are malicious but metasystems have no ethics or concern for human welfare and the metasystem of the Mind Benders must be considered at least a potential threat to humanity.

As with all metasystems I see the mind-benders as a natural development rather than a Machiavellian plot but, because it can actually change human nature, I consider it very dangerous.

I don't suggest that all Mind Benders are malicious but, because advertisers support the media of their choice, media that reflects their views prosper and non-conformers do not. Most people are not aware of the triumph of the mind-benders but they suffer from it, because the best way to sell something is to make people unhappy with what they already have.

Since the development of the cash economy, money has ruled the world and, because they control advertising budgets, the mind-benders hold veto power over the media and they have molded much of it to suit their purpose.

I used to work for establishment media myself, and when radicals told me how the media serves the establishment I knew better, because I knew — and still know — that I wrote what I believed.

But advertisers support the media of their choice, so media that reflects the views of the advertisers prospers and others do not. That's a problem because the best way to sell something is to make people unhappy with what they already have.

If I want to sell you a new car I must first convince you that your old car — or whatever form of transportation you're using now — is not adequate for your needs.

It does me no good to prove that the Drof car I am trying to sell is better than the Egdod that someone else is trying to sell if I have not already convinced you that you need a new car.

In fact salesmen for both companies try to convince you that you need a new car and, because they pay for the content of most mass media, the media itself reinforces the message. Because we are taught that having a new car is a desirable norm, the self-image of people who can't afford one is debased.

As a salesman I may spend more to have some 'responsible' journalist spread my message than for advertising, because you as a reader may be suspicious of advertising but will accept the journalist's comments as an honest judgment.

That's why some advertisements are disguised to look like news stories.

As a former journalist myself I can assure you that most journalists do in fact print their honest opinions, but that doesn't mean much. The control process is informal, but it leads back to the mind-benders.

Advertisers put their money where they think it will do the most good, so they support media that appeals to people who may buy their products. If I advertise Drof cars, I will use media I think will appeal to people who might buy Drof cars.

If the media looks good to me that's because the publisher shares my view of the world, and he hires editors and reporters who share his view of the world. There is no plot, but because advertisers hold the purse strings, they decide which media will prosper and therefore which view of the world will be disseminated.

Chain Media

The power of the Mind Benders has been consolidated and multiplied with the development of chain media.

Fifty years ago local newspapers and radio stations were locally owned and they reflected local opinions. Now most of our newspapers, magazines and radio and TV stations are owned by giant corporations and they reflect and support opinions that benefit those corporations. The newsrooms of some newspapers have been merged with the newsrooms of TV stations and, for a while, newspapers in one Canadian chain were required to publish editorials that were written at the chain's head office.[235] That policy was changed, under public pressure, but it's hard to believe that any newspaper is free to express opinions that are not approved by its owners. One way or another, the company that pays the editor controls the news.

The take-over of local newspapers by chains made a big difference because many of the owners of independent newspapers believed in public service and were willing to stand up for their beliefs. Some of the people who control newspaper chains may have an active interest in manipulating the government and the public and many of the managers of individual papers in a chain are bean counters who are judged only by the profits they can wring out of their publications.

Governments probably did not plan to help media moguls to build chains of newspapers, but they did help. Roy Thomson, who built one of the world's big media empires, said that family-owned newspapers are not economic in Canada because the heirs have to pay inheritance taxes every time the property is passed on.[236] American chains were helped by an IRS ruling that the cost of buying newspaper companies was a legitimate business expense. Because of that, a chain that invests its profits in the purchase of more newspapers can avoid taxes.[237]

Now we live in a sea of advertising and propaganda and I do not suggest that we could do without it, but I consider it vital that we understand it. The people who create and manage propaganda and advertising are even more vulnerable than the rest of us because they are immersed in it. As Jacques Ellul notes, people who can't read and don't have radios are not exposed to propaganda and are therefore not affected by it.

The effect of propaganda is intensified if all the media offer the same news and the same interpretation of news and, in North America, news has become more standardized over the years. The trend began with the rise of syndicated columnists who present the same opinion to millions of readers at a time and the standardization has become more complete with the growth of chain media that presents the same picture — as approved by the management of the chain — to readers across the country.

Both advertising and propaganda are legitimate — even necessary — activities but they can threaten society because they can drive civilized people to un-civilized behavior. Every war we know about provides abundant examples.

Modern Mind Benders are a threat to progress because big companies know that it is more profitable to invest in advertising than

---

235 "Their Master's Voice," *The Globe and Mail*, Dec 10/01, pg. A18.

236 Braddon, Russell, *Roy Thomson of Fleet Street*, Collins, London, Toronto, 1965, pg. 148.

237 Bagdikian, Ben, *The Media Monopoly*, Beacon Press, Boston, 1983, pg. 10.

in research and development. According to one source American drug companies now spend twice as much on advertising as on research.[238] Another reports that if all marketing efforts are included, the total is closer to 19 times the amount spent on research.[239]

We see one side-effect of the reliance on propaganda and advertising in the decline of the American automobile industry. Once the biggest and most advanced in the world it is now an also-ran, largely because American auto-makers concentrated on advertising while European, Japanese and even Korean auto-makers concentrated on developing better cars.

Communication is the key to social life and propaganda, advertising and media are all forms of communication. Our society could not survive without them but if they run amok — as seems to be happening — there may be a question of whether we can survive with them. As a former reporter I believe in free speech and 'freedom of the press' but I also know that no long-lived eusocial society has anything like them. I don't think we can actually make a choice but I suspect that if humanity survives long enough, evolution will make a choice for us.

Between them religion, advertising and propaganda — and now social media — have taken the power to create conventional wisdom away from the elders of our society and given it to people who use it for their own benefit. More, because mass media dominates our society, it spreads the same conventional wisdom through all castes and classes.

Cell phones, the internet and social media are expanding the metasystem of the Mind Benders exponentially and I won't try to predict the end result but I suggest that, by giving us species-wide communications, they all contribute to our eusociality.

In the presidential election of 2016 Donald Trump was unpopular with his party and with the media but he used the power of social media to gain the Republican nomination and win the election.

Trump himself insists that 'false information' is becoming ever more prevalent and that's certainly true because social media, cell phones and the internet allow anyone to disseminate ideas and information — true or false — without control. That's kind of ironic because Trump's electoral campaign was based mostly on social media and, personally, I think it might be interesting if some expert in subliminal messaging were to analyze all Trump's 'tweets' and other promotional materials.

Some of the information spread over 'social media' — like fake 'news' stories about Sharia law being adopted by some American cities and states — are deliberately divisive. In response to one such report a

[238] https://www.sciencedaily.com/releases/2008/01/080105140107.htm

[239] http://www.huffingtonpost.com/2012/08/09/pharmaceutical-companies-marketing_n_1760380.html. Cites the *British Medical Journal*, http://www.bmj.com/content/345/bmj.e4348

half-baked vigilante drove from North Carolina to Washington DC to shoot up a family pizzeria that had been described as the headquarters of a child-sex ring.

In the wake of Hurricane Harvey, which flooded several cities in Texas, one social-media 'news report' that a mosque was refusing to help Christians was re-tweeted at least 200,000 times. The mosque named in the report does not exist, the address given does not exist, the imam named in the report does not exist and the photograph of him, included in the report, is of the imam of a Toronto mosque who has never been in Texas. I don't know whether any American media reported the real facts, but the fraud was reported on a national CBC newscast in Canada and in the *Toronto Star* newspaper.[240]

The War on the Family

For most of the past century the Mind Benders have mounted a deliberate and sustained attack on traditional families. That's a serious charge but I don't argue that the advertisers who promoted it were consciously malicious. Rather, I suggest that they wanted to make a living and did not think beyond their own interests.

The attack on the family began soon after the First World War, when industries that had expanded to meet wartime needs found that they had more productive capacity than the peacetime economy could absorb. The obvious solution was to sell more consumer goods but that would not be easy because in those days many Americans — especially the farmers who were still the backbone of the country — were nearly self-sufficient. The key, advertising psychologists realized, was to transfer control of the family from the adults to the children.

In *Captains of Consciousness* writer and professor Stuart Ewan quotes several industry leaders on the need to teach children to reject their parents' standards.[241] He quotes psychologist Alfred Poffenberger on the need to advertise directly to children because adults have habits that are hard to break but children can be counted on a demand change. About the advertising of the era Ewen writes:

> ...the ads often painted a picture of adults as incompetent in coping with modernity, and raise the model of youth as a conduit to consumption. A 1922 ad from Paramount Studios documented the breakdown of parental know-how and gives an insight into the ways in which youth served as a cover for new authorities. Reinforcing the need for "keeping up with youngsters," the ad informed readers that "the young folks do their parents every bit as much good as their parents do them."[242]

[240]"Photo of GTA imam used for fake news," *Toronto Star*, Mon Sept 4, p A7.

[241] Ewen, Stuart, *Captains of Consciousness*, McGraw Hill, New York, 1976, 1977 paperback edition, pp 144 -149.

[242] Ibid., pg. 147.

One of the leaders of the movement was John B. Watson, a professor of psychology at Johns Hopkins University and one of the founders of behavioral psychology, who considered all but the "gratifications of the marketplace" to be perverse and socially damaging.[243]

In 1920 Watson left Johns Hopkins and later went to work for the J. Walter Thompson advertising agency,[244] and much of his work reflects the kind of personality that seems to fit better with an aggressive — even predatory — career in advertising than with a career as a university professor. In one famous experiment he conditioned an 11-month-old boy to fear white rats and other furry objects.[245] In a book on child care he advised parents to "never hug and kiss" (children) "never let them sit on your lap. If you must, kiss them once on the forehead when they say good night. Shake hands with them in the morning. Give them a pat on the head if they have made an extraordinarily good job of a difficult task."[246]

Other writers took up the cause. In 1931 Edward Filene, a department store magnate (and founder of the credit union movement in the United States), argued that in the modern world the "head of the family is no longer in control of the economic process through which the family must get its living," and that, rather than look to a husband or father for guidance, women and children must look to "the truths which science is discovering."[247]

This was not a malicious plot against society. It was just a case of people doing what was best for them and for the companies they served. If they had been more concerned about the welfare of the family and of society than about keeping their jobs, they would have taken a different course (and probably failed), while someone else wrote the ads that attacked the family.

The collapse of the traditional family in the Western world may have reached a tipping point in the 1960s with the rise of The Beatles and their new and overwhelmingly popular 'psychedelic' music, which tied into the promise of the new psychedelic drug LSD. This also helped popularize marijuana, which has been a popular drug for thousands of years.

The American government had been trying to stop the use of marijuana for years but, with The Beatles and LSD on one side and a new distrust of the government on the other, American youth embraced it.

The war itself did not have much impact on America but the lies did, and the anti-war protest movement was born. This was

243 Ibid., pg. 83.

244 https://en.wikipedia.org/wiki/John_B._Watson

245 *Encyclopaedia Britannica CD 98*, John B. Watson.

246 Watson, John B, *Psychological care of infant and child* WW Norton, 1928. My quote is from page 81 in the 1972 reprint by Arno Press & The New York Times, which is based on a copy of the original from the University of Illinois library.

247 Filene, Edward, *Successful Living in this Machine Age*, Simon and Schuster, NY, 1931, pg. 96, quoted by Stuart Ewen in *Captains of Consciousness* pg. 82-83.

significant because it produced an explosion of propaganda and widespread dissemination of propaganda techniques.

The protest movement included a lot of psychologists who knew how to develop propaganda and who were now in a position to experiment and develop new techniques. The counter-system of the anti-war movement forced the Military Industrial Complex to abandon the Vietnam war and, incidentally, produced a quantum leap in the science of mind-bending.

Between them, the war and the anti-war movement created a rift in the American population and an opportunity for the development of a 'youth culture' that estranges teens from their parents' values and encourages them to accept ideas and standards only from the agents that developed the youth culture.

In the 1970s, one hit song featured a teenager's plea, "This is my life. Go ahead with your own life and leave me alone."

More than fifty years ago Austrian naturalist and psychologist Konrad Lorenz argued that the modern antipathy of the young for the old goes beyond normal limits. It's not a conflict between two segments of one culture, because modern teens are conditioned to see themselves as an entirely different nation from their elders. In *Civilized Man's Eight Deadly Sins*, he wrote:

> The attitude of many of the younger generation toward their parents shows a good measure of conceited contempt but no understanding. The revolt of modern youth is founded on hatred, a hatred closely related to an emotion that is most dangerous and difficult to overcome: national hatred. In other words today's rebellious youth reacts to the older generation in the same way that a cultural group or 'ethnic' group reacts to a foreign, hostile one.[248]

Lorenz wrote that in the early 1970s, in a small town in Austria — a time and place when the youth culture was relatively low key. If he could see modern North American culture, he might say that the split between young and old may be terminal.

Apologists for youth suggest that we of the older generation can't understand the younger because we can't deal with the pace of technological change in the modern world. In fact, most of the younger generation has never seen a significant change in technology.

My father was born in 1887 and he died in 1965. When he was young, street-cars were pulled by horses, there were no cars in the city where he lived, telephones and electric light were experimental and aircraft were a crazy dream. Before he died, he watched color television and travelled by jet plane. He adapted to change on a scale that modern youth cannot imagine.

[248] Lorenz, Konrad, *Civilized Man's Eight Deadly Sins*, translation by Marjorie Latzke published in London by Methuen & Co. Ltd, 1974, pg. 48.

On the other hand, modern youth lives in a social culture that older generations find hard to believe. When my father was young, any North American man with a few years of grade school education could expect to find a secure job, to be able to support a wife in his own home and to support his children until they finished high school.

Every young woman could expect to marry and to be supported in her own home while she raised her children. If her husband left her, she would still be entitled to support for the rest of her life or until she remarried and, if her husband died, her family would help raise the children

When I was young our expectations were about the same, with the exception that it was quite reasonable for an average man to expect to pay for his children's university education.

In 1967, I was a reporter for a small city newspaper — a job that did not demand a university degree in those days. My wife did not work and I bought a three-bedroom house, with 160 feet of river frontage, within a city, for slightly less than one year's salary. The water was clean enough to swim in or to drink, for that matter. Because of a dam downstream about a mile of river behind my house was more like a lake than a river and, for a while, a friend kept his small sailboat at my back door.

Now the average man with a university degree knows that he will be lucky to find a job; that if he does the job will not be secure; that whether his job is secure or not his wife will probably have to work, that they may never own their own home and that their children will likely have to borrow money to pay for a university education.

The physical changes in my time have been relatively minor. My car is more efficient and dependable than my father's was, but it's not all that different. Some cars that were built before I was born are still running and the electric cars that some visionaries see as the ideal were more popular 100 years ago than they are now or are likely to be in the foreseeable future. A few years ago TV commercials advertised "the first truck in the world with four-wheel steering." It must have been an old one, because the British, American and Russian armies used Nash 'Quad' trucks with four-wheel drive and four-wheel steering in the First World War.

The first car I owned was made in 1937 and, given a choice, I would rather buy that car new than many modern ones. The first airliner I flew in had propellers, but it may still be flying today, and within the past thirty years I have flown in planes that were designed before I was born. Television was not common when I was young but the Berlin Olympic games of 1937 were televised to 25 'viewing rooms' around the city, and the first paid TV commercial was broadcast on the afternoon of July 1, 1941, over New York station WNBT (later WNBC) before a baseball game between the Brooklyn Dodgers and Philadelphia Phillies.

The biggest physical change in my lifetime is the development of computers, which affects me because I use one, but the work I do on a computer is essentially the same as the work I used to do on a typewriter. About the time my father was born writers were giving up pens for typewriters, which was an equivalent change.

Computers represent a bigger change for accountants and some others than for me and, for most of us, their most important effect is the delivery of propaganda and the kind of information that Donald Trump calls 'fake news.' Computer chips make our watches keep better time and our cars run better, but the old watches and the old cars were good enough for most purposes. Some computers monitor and control industrial processes and they do good work, but they also displace people and, for most of us, their most important effect is to create unemployment. In this respect they are a continuation of a process that began a couple of hundred years ago.

Some men now work in space, but they are not much more real to the rest of us — or more relevant to our daily lives — than were the science-fiction heroes of fifty years ago.

Orbiting satellites are useful but, while the exploration of outer space is a gravy train for some government departments and the companies that collaborate with them, for the average human being it's mostly a waste of tax dollars. I gained no benefit when men landed on the moon and I expect none if men land on Mars. One so-called 'philanthropist' is promoting a colony on Mars, but the fact is that the cost of one person's travel to Mars would keep a family in luxury for a lifetime on Earth. The only possible justification for a colony on Mars would be if it enabled some humans to survive if Earth were so badly damaged that it could not support life — but the creation and support of that colony would require the construction and use of hundreds of big rockets and thus contribute to the destruction of Earth's capacity to support life.

The most important changes in the modern world are in our social structure, and they are not improvements. When my father was born the root systems of armies, churches, the state and so forth existed but, except for the schools and churches, they had little influence on most of us. By the time I was born in 1937, we had plenty of both machines and systems, but the Mind Benders had little power and neither machines nor systems ruled our lives. During World War II, my older brother and I used to listen to a radio serial about the crew of a Lancaster bomber called L for Lanky, but that was for a half hour every Sunday. The rest of the time we played with other children and the dominant influences in our lives were our parents, our friends and our teachers. My brother had a record player but Bing Crosby, Frank Sinatra and other singers of the day sang about the same values my parents held.

The biggest changes I have seen in my lifetime are the destruction of the family and the increase in poverty. The triumph of the youth culture is a boon to business because the beliefs and standards of the older generation evolved before the development of powerful propaganda and advertising techniques, and too many of them include human values. Like the youth culture itself the values and standards of the youth culture have been established by the mind-benders and, by instilling those values in modern youth, the mind-benders gained total control of a whole generation of humanity.

Youth in Rebellion

Apologists for the youth culture suggest that youth has always been rebellious. That is partly true, but the rebellion of the past is not to be compared with the rebellion we see today. This is more like civil war than rebellion and it's even more tragic than other civil wars because adults are predisposed to yield, where possible, to the demands of children. When youth rebels with determination it is sure to win — but if it wins the wrong battles it will lose in the long run.

The prophets of the new world are pop singers and their bands, and many of the top singers are children themselves. Typical teen bands do not present an adult-oriented view of the world. Some of them have real power, but their success depends on promotion by Mind Benders and the ones that are chosen for success are the ones that set a tone that suits the Mind Benders.

The substance of modern teen music is that the highest aim for a human is to renounce family ties and to join some system-based group — plus whatever shtick the figurehead chooses as his or her personal signature.

In a world of teens who have learned to resent adults, anger and hatred are good signatures. If Satan's Sodomists or some other band spews hatred and resentment it may not sell records to adults but, whatever they think of the music, some teens will be attracted to any band that adults dislike. Peer pressure will attract others and, because adults dislike it, the band and the attitudes it promotes can be a hit.

Music groups may foster and exploit resentment but they always direct their venom against authority, never against the Mind Benders who control them.

The Battle of the Sexes

In the 1970s the Mind Benders expanded their 'War on the Family' with the feminist movement, which encourages women to relate to each other rather than to husbands and children. The roots of the movement date back hundreds of years — the English philosopher John Stuart Mill wrote about the oppression of women by men — but modern feminism got a boost with the publication of a book. *The Population Bomb*, by biologist Paul Ehrlich, painted a horrifying picture of a world that would soon be hopelessly overcrowded.

In his prologue Ehrlich says, "The battle to feed all humanity is over. In the 1970s the world will undergo famines — hundreds of millions of people are going to starve to death in spite of any crash programs embarked upon now. At this late date nothing can prevent a substantial increase in the world death rate."[249]

Ehrlich speculated on the world in some unspecified future when the population might be sixty quintillion people — nearly ten million times as many people as there are now. In this horrifying scenario the whole planet would be covered by a single building 2,000 stories high, in which each person would have about three or four square yards of floor space.[250]

The book was very successful and it helped to promote the new oral contraceptives that had been introduced in 1960. Many people saw birth control as the only possible cure for the predicted catastrophic overpopulation and, with techniques that had been perfected to protest the Vietnam war, activists began to protest against motherhood and the traditional role of women.

If they were 'freed from the bonds of motherhood' women could take full time jobs, for which they would be paid in cash. More, if they abandoned their natural role they would be more easily manipulated into the new roles that the mind-benders planned for them. This sounds Machiavellian and it is, but I do not insist that it was deliberate. Rather, it is an example of the phenomenon that I call *drift*, in which trends that are good for a metasystem are encouraged by all members of that metasystem.

On one level, working women were not much of a change because women have always worked. In the early days of humanity our male ancestors hunted and our female ancestors gathered food, and the gathering usually provided more (and more dependable) food than the hunting did.

Even among Mongol herdsmen and the Inuit, both of whom are mostly meat eaters, the diet includes roots, berries and teas or soups made of bark, lichen, berries and grasses — all of which are gathered rather than hunted.

In most tribal societies women gather wild food and plant and manage gardens. In wheat-growing cultures, men worked the fields while women cleaned and ground grain and baked bread. In most pre-industrial societies, women also dress hides, collect fibers, weave cloth and make clothing.

In industrial times women made, washed and repaired clothes, baked bread and prepared other foods. Through most of human history, the home has been a primary center of production.

To people who believe that everything should be made in a factory that may sound mediaeval, but work at home does not have to be unpleasant. In the late 1990s I spent six weeks as the guest of an Inuit

[249] Ehrlich, Paul, *The Population Bomb*, Ballantine Books, New York, 1968, prologue.
[250] Ibid., pp. 18-19.

family in the village of Rankin Inlet, on Hudson's Bay. My host was well-off and the most comfortable room in his house was the sewing room where his wife and her friends made and repaired clothes for their families. It had about 600 square feet of floor space and, in addition to a couple of commercial sewing machines, a serger and a big cutting table, it had a comfortable couch and some easy chairs, a 36-inch TV set with satellite dish and VCR, a kitchenette and a bathroom. The sewing room was a workshop, but it was also a very comfortable social center.

Most of the women of Rankin Inlet make most of their own and their families' clothes, because they can make better clothes than they can buy. When my mother was a child, early last century, her family never bought women's clothes. Instead a professional seamstress came to the house for one month each year to help my grandmother and her six daughters make everything they would need for the year.

When I was growing up in the 1940s, my mother still made most of her own clothes, most of my father's shirts and many of her children's clothes. The clothes she made were much better quality than we could buy and they were worth a considerable amount of money.

My mother also made most of the family meals from scratch and, during the Second World War, she knitted socks for soldiers and grew much of the family food in a 'Victory Garden.' We were not poor; but that was how people lived in those days.

Most of the women in the neighborhood were home every day, most of them were friends and they often visited back and forth. The men were not all friends but they all knew one another and, essentially, everyone in the neighborhood knew everyone else.

I think my mother had a good life in those days — but in a world threatened by overpopulation, motherhood and housekeeping are now discouraged. Much better, the feminists say, for women to go to work, and if they must have children, let the schools and day care centers take care of them.

Whatever the effect on families, women's liberation looks good for the economy because women who work at paid jobs are more likely to buy clothes than to make them and to buy prepared food and restaurant meals rather than cook their own. Working women need more clothes, cars and services than women who stay home, thus creating more consumption for commercial businesses and more taxes for the systems of government.

Women who do not marry or who leave their husbands also increase the need for living space, because men and women who need only one house or apartment as a couple need two if they live apart. In some cases children are abandoned by both parents to live in government-supported foster homes.

Working women also enlarge the labor force and, in a time of declining real production, contribute to unemployment — which helps keep wages down. Overall, feminism has been a good career for the women who promote it, but it's a questionable benefit to most women, and to humanity as a whole.

The women's movement is another big step toward eusociality, because females are the core and sometimes the only members of a eusocial society. Among ants and honeybees there are few males, and they live only to mate once before they die. Termite colonies support about as many males as females, but most of the males, called pseudergates, are neuters — as are most of the females.

Among naked mole rats only the queen breeds, and she may have several mates, but most of the other members of the colony are effectively neuters. The male breeders can be identified because they have visible testes, but most males do not.[251]

Meerkats live in 'mobs' of 20 or more, but only one pair is allowed to breed. If another pair breeds, the dominant pair will usually kill the offspring.[252]

Herds too may be matriarchal. A herd of elephants is usually made up of closely related adult females, led by the oldest female, with their offspring of various ages. The basic herd is usually small, but in the dry season two or more herds may join to form a larger bond group, and bond groups may unite to form a clan. These larger groups are not as close as herds, but their members will cooperate to defend territory. Adult males join herds when there is a female in heat and adolescent males leave herds to form small bachelor groups, which are loosely structured and unstable.[253]

Among most species, females are more cooperative and easier to lead or manage than males. Mares are easier to ride than stallions, cows and ewes are easier to herd than bulls or rams, and so forth. We might guess that cooperation and yielding to requests are survival positive for females, because most of the requests they receive are from their young, and if they don't respond, their young may not survive.

In the modern world women are encouraged to compete, rather than cooperate. This is a big change for humans, but it brings us into line with the ants, honeybees and other eusocial species.

Women's Work

Women have always worked. in hunting and gathering cultures they gather food, dress hides, collect fibers, weave cloth and make clothing.

Even in early industrial times they made, washed and repaired clothes, baked bread and prepared other foods. In rural and semi-rural areas they often had gardens, and even in cities they might keep hens or cattle.

251 https://en.wikipedia.org/wiki/Naked_mole-rat

252 https://en.wikipedia.org/wiki/Meerkat

253 http://www.africam.com/wildlife/elephant_herd_structure. See also https://en.wikipedia.org/wiki/Elephant#Social_organisation

Through most of history they also had an active social life because they gathered food, tended gardens and did other work in groups. The big change came with the industrial revolution because, among other side effects, it created the nuclear family.

Up to this point most people lived in extended families, which often included three or more generations and two or more married couples because a farmer's son does not have to leave home when he marries. If they moved to a village they could still live close together, sometimes in a family compound — but the extended families broke up when some members moved to towns to take factory jobs. Even if they all moved to the same town there was no housing suitable for extended families, and they couldn't live as a group.

Most nuclear families were dominated by a man but women contributed to the family income, and many of the first factories were spinning and weaving mills where women made cloth.

In World War I English and North American women went into other factories as welders, machinists and other 'men's jobs.' That gave them a different outlook on life and in the 1920s and 30s women — encouraged by the Mind Benders' war on the family — celebrated their new freedom as 'flappers,' often with careers.

They went back to the factories in World War II but, when economists predicted a return to the Depression after the war, most of the women went back home and left the factories to men. The combination of rising production and consumption produced a boom that lasted for nearly 20 years, but it also spawned the phenomenon of suburbia and some of the worst social and living conditions women had known for centuries.

Suburbia was supposed to be good for women and at first glance it looked good, but in some ways the suburban housewife became little more than another household appliance.

In earlier times and still, in other parts of the world, urban women went shopping every day. Each time, they would walk through a village to the butcher, the baker and the fruit and vegetables markets, meeting and talking with people all the way.

They would personally choose the food their family would eat, almost the way their ancestors gathered the best roots and berries, and they might bargain with the shop-keeper for the best price.

Because most of their friends also shopped every day, the shopping trip was a friendly social outing. They did their shopping close to home and most of the women knew most of the shop-keepers and most of the other shoppers.

Modern suburbia changed that. With freezers and refrigerators to store food and cars and station wagons to carry it, a suburban housewife need shop only once a week, and she often drives miles to a supermarket to do it.

When she shops on a village street, the people she meets are friends to talk to, but as she drives to a supermarket the drivers of other cars are strangers and, because of the danger of an accident, potential threats to her life.

She parks her impersonal car in an impersonal parking lot and does her shopping in an impersonal supermarket where she does not know the store-keeper, the staff or the other shoppers. She has little chance to inspect the food she buys because much of it is pre-packaged, and she cannot bargain for the price.

Her shopping for the week is done in a few hours but, instead of being a pleasant social experience, it is a potentially dangerous trip into a foreign and unfriendly world. When she gets home she can play bridge with her friends, but you can only play so much bridge.

She could read but, as Betty Friedan points out in *The Feminine Mystique*, the women's magazines of the day went through a radical change.[254]

In the 1930s and 40s, the best women's magazines were excellent journalism with a mix of fiction, articles about career women and some home-oriented service articles. In the 1950s they began to concentrate on 'service' articles and housewives.

Herself a writer for women's magazines, Friedan blames the change in magazines on the change from mostly-women to mostly-men writers and editors — but she misses the point. She admits than women worked cheaper than men, so the magazines didn't save money with the change.

They had to change because a metasystem was forming. Some people call it 'women's liberation' but I consider that a misnomer for a movement that takes women out of the home — where they are essentially autonomous and need consider only their own and their family's interests— and puts them to work at jobs that are of no direct interest to them under supervision by people they have no other relationship with.

Spurred by Ehrlich's dystopic view of an overcrowded future and the Mind Benders' propaganda and advertising –- and the anti-human conditions of modern society — many women, some of whom were already married, gave up the idea of women's traditional role and decided on a life style that to could be compared to that of a pseudergate in a termitary. Whether this is good for women or not, it offers wonderful benefits to a wide range of commercial interests.

This is easier to understand if we recall that metasystems form when the behavior of one entity benefits, threatens or otherwise affects the behavior of others. The 'women's movement' is a metasystem in which the members all do what's best for them and the sum of their behavior — whether beneficial to humanity or not — is good for the metasystem. In this case, many of the beneficiaries are commercial interests.

Working women are more likely to buy clothes than to make them and to buy prepared food and restaurant meals rather than cook their

[254] Friedan, Betty, *The Feminine Mystique*, Norton, NY, 1963.

own. Further, working women need more clothes, cars and services than women who stay home, thus creating more consumption for the business community and more taxes for government.

As women were encouraged to not marry, or to break up existing marriages, they even increased the need for living space because men and women who need only one apartment as a couple need two if they live apart.

Lawyers, real estate agents, 'therapists' and psychiatrists benefited from the break-up of families, there were more jobs for social and child care workers and other government employees and a whole new industry of day-care for children was established. Partly because mothers who did not care for their own kids felt guilty and partly because kids who were not cared for by their mothers were more demanding, sales of toys soared. Kids on a farm or in a craft village can make their own fun but kids in a manufacturing culture need manufactured toys.

Business also benefited in several ways from the 'meat rack' bars that opened as meeting places for singles. Besides business for the bars themselves the 'bar scene' created more demand from both men and women for expensive clothes and impressive cars and jewelry. The mating period, when both men and women do their best to look attractive, was extended indefinitely.

I'm not suggesting conspiracy here because, as noted earlier, metasystems are self-organized and most of the time most people will take the stance that offers them the most benefits. It's perfectly natural and understandable — even justifiable — that, for example, when more women began looking for work outside the home women's clothing stores would feature more 'business' clothing. When marriages break up divorce lawyers prosper and, if people live as singles, more houses or apartments are required. On the other side of the coin, the more businesses that cater to singles, the more attractive single life looks.

It was also a benefit to business interests that the flood of working women put a cap on wages. Whether they were paid as much as men or not, working women increased the size of the work force, thus creating unemployment and a lever to keep all wages down. The average North American has lost about 20 percent of real purchasing power since the mid-1970s.

Official numbers show that the average pay of women working full-time for a full year increased from 60 percent to 72 percent of the pay of comparable men between 1975 and 1997, but much of that was due to the loss of men's wages. The real annual earnings of women rose by just 12.8 percent, even though they worked more hours. Even now, only half of all women in the labor force work on a full-time, full year basis. Between 1983 and 1992, the median real annual earnings of young women workers aged 20 to 24 fell by 24 percent and those of young men by 22 percent.[255]

There were some gains for men at the top — 6.2 percent for the top 10 percent — but the real annual earnings of the bottom 90 percent of men fell and the bottom 10 percent of men lost 31.7 percent. This drove more women to work outside the home, of course, because the average man's wage could no longer support a family.

You might think that more workers would produce more goods and that bringing women into the work force would increase production, but it didn't work that way. The women's movement argued that women could work like men, but the women who went to work didn't want jobs in factories or mines or on construction. They wanted to be lawyers or media personalities or bureaucrats.

Text-book economics considers any paid work to be 'productive' but, in real life, productive work is work that increases the wealth of society as a whole. Work that distributes wealth produced by others may be satisfying for the people who gain the wealth but it is not productive. The flood of women going to 'work' increased consumption, but it did not increase production.

It did, however, increase production in third world countries — and contribute to the United States' now-unmanageable debt — because few first-world women want to work in factories and many of the clothes and other goods that 'working women' need are imported. This effect was compounded when cheap goods made by cheap labor forced many first-world factories to close.

Under populist pressure some governments collaborated with the women's movement. In the 1990s, for example, the then-socialist government of Ontario created 1,500 new jobs it called "bridging positions." These were frankly make-work jobs, and their sole purpose was to put women into 'executive' positions. Given that the first job of a civil servant is to find something to be seen doing, most of these jobs soon appeared to be an integral part of the machine of government.

It's a safe bet that by now most of the original 'bridging positions' are established jobs and that the people who fill them have private offices, secretaries and assistants.

The women's movement is supported by the media and the advertising industries partly, at least, because women are more vulnerable to advertising than men are. This is a natural development because in most species females cooperate while males compete.

Nature can't afford to experiment with females because everyone must breed to maintain the species. Males are the wild card that drives evolution because one male can fertilize many females. Nature can afford to experiment and let the winners breed.

[255]https://www.policyalternatives.ca/publications/reports/falling-behind#sthash.dQofzjRF.dpuf

And in some ways, nature may be taking a hand in breeding selection. Researchers at the Hebrew University of Jerusalem who studied data from 185 studies of 43,000 men conducted between 1973 and 2011 found a 52.4 percent decline in sperm concentration and a 59.3 percent decline in total sperm count in men from North America, Europe, Australia and New Zealand. This is believed to be caused by a combination of air and water pollution, artificial chemicals in our diets and/or our clothing and environment plus smoking, drinking and tension. Most of these factors are environmental but we have more control over our environment than any other animal and, in theory at least, we should be able to change them. Unfortunately, metasystems have more control than we have.

No significant decline in men's virility has been found among men in South America, Africa and Asia. This might be because fewer studies have been done there, or it might be because metasystems in less developed countries are not as strong as they are the Western World.

The Mind Benders seem to have more power over women than over men, but that is at least partly because most advertising is aimed at women.

At any given time, in Canada at least, there are several TV commercials that show men as liars, unfaithful and/or incompetent in contrast to women who are honest and competent. These commercials are good business because women are more easily persuaded than men, and men have no militant metasystem to defend their image. Any commercial that insulted or denigrated women would be actively protested by women's organizations and their male supporters alike.

As I write this Buick cars boast that owners can lock or unlock them with a cellphone. In one TV commercial a man and a woman are lying on a beach. The man assures the woman that he remembered to lock the car, as he surreptitiously locks it with his cell-phone. He then says that he remembered to close the windows in their apartment and the video cuts to a view of an apartment, with windows open, being invaded by a flock of pigeons.

What's the real message in that commercial? Is it that Buick cars can be locked by cellphone, or is it that men are liars and not dependable?

In one Ontario Motor League commercial a man who has locked his keys in the car throws a hissy-fit while his wife calmly phones the Motor League. Some patent medicine commercials fake a class in medical school with dumb men and bright women students, and so forth.

Advertisers may be right when they assume that anything that disparages men with the stereotypes that appeal to man-haters will sell to women, but the effect goes beyond sales. If men are nagged long enough, they may react with resentment.

Most men will not resort to deliberate violence but the cost of tension is buried in other figures. Because a driver tries to work his tension out on the road, two cars collide, and perhaps someone is killed. Safety officials call that "road rage" and they estimate that it now kills about 200 Americans a year — an increase of 51 percent since 1990. An Australian study estimates that road rage causes about half of all traffic accidents.

Rage causes other problems, too. Because someone is tense and consumed with hate a machine is not serviced properly and it breaks down. Because someone is tense a marriage breaks up, and a child who might have been productive grows up to be a drain on society.

And because of the hate campaign we are now more likely to see fellow citizens as enemies than as friends. Many men now do not expect even common courtesy from a strange woman and, because they don't expect it, they may not offer it.

It's a brave new world we live in, but it's not one that many people like. Even the hate-mongers who made it and who profit from it probably don't like it, but perhaps they may be victims of their own propaganda.

Much of the fury has died out of the women's movement now but the furies still have power and men have memories of what women call a "poisoned environment."

Men's rage has been simmering for 50 years, and some women pay the price but, as we might expect, the women who do most to create the hatred are well insulated from it with comfortable sinecures and fat bank accounts.

But hatred and rage are now major factors in our economy and our politics. They are good business for some but a disaster for the economy as a whole. Hate-mongers are hard to fight, partly because so many of them are victims of their own propaganda, and they actually believe that they are right.

Many of the people who run hate campaigns know enough about human nature that they should understand what they are doing, but they seem to be blinded by hatred, self-interest and their own propaganda.

A Somali proverb says "you cannot wake a man who is pretending to be asleep." One assumes that it would be even harder to wake a man who has found a way to be paid for sleeping.

The Plight of the Children

Men and women both suffer from the Mind Benders' war on the family but the real victims are the children and, perhaps, our future. In the 1950s, 80 percent of North American children lived in homes with two biological parents who were married to each other. By the end of the 1980s, only twelve percent of children lived with two biological parents. Psychologist Shirley Steinberg cites studies that found that more than half of all North American children's parents are divorced and that children of divorced parents are almost three times as likely as children raised in two-parent homes to suffer emotional and behavioral difficulties.[256] We might also assume that they are less likely to have stable marriages themselves, and that their children will have more problems than they have.

Broken homes and re-marriage create a serious human problem. Government figures on child abuse and child mortality don't discriminate between natural and step-children, but McMaster University psychologists Martin Daly and Margo Wilson have extracted disturbing numbers from other data.

They found that in Hamilton, Ontario, a step-child is about forty times more likely to be abused than a child living with both natural parents. Across North America, they estimate, a child living with substitute parents in 1976 was about one hundred times more likely to be fatally abused than a child living with both natural parents.[257]

I won't list the horrifying examples of this that we have seen in Canada in the past few years, but a google search for "children killed by substitute parents" turned up more than 4 million results.[258]

Child abuse and violence may be partly a cultural inheritance. Dr. Terrie Moffitt, associate director of the Dunedin Longitudinal Study, which follows the life history of 1037 people born in 1972 in New Zealand, says the study found that all the subjects who were violent and antisocial as adults had two characteristics in common. One is that they were lacking, or had a weak form of, the gene MAOA, The other is that they all have a history of maltreatment or neglect before age 11.[259]

About 30 percent of people lack the gene or have a weak version of it, but this trait apparently has to be combined with neglect to produce anti-social behavior. Some people who lack the gene but were raised by good parents are more successful than average, and they are valuable members of society.

Archaeological evidence suggests that child abuse in any form is a modern development. Anthropologist Phillip Walker has examined the bones of more than 5,000 children from preindustrial cultures

---

[256] Steinberg, Shirley R., "Kinderculture: mediating, simulacralizing and pathologizing the new childhood" in *Kinderculture*, Shirley R. Steinberg ed, Westview Press, Boulder, ColO, 2011, pg. 3-4.

[257] Daly, Martin and Margo Wilson, *Homicide*, Aldine de Gruyter, Hawthorne, NY, 1988, pp. 88-89.

[258] http://www.google.ca/search?qchildren+killed+by+substitute+parents&hlen-CA&gbv2&oqCHILDREN+killed+by+subtitute+parents&gs_lheirloom-serp.1.0.30i10.97952.108784.0.110506.11.11.0.0.0.0.109.889.10j1.11.0....0...1ac.1.34.heirloom-serp..9.2.177.w-yx8rMYLvY

[259] film, "*When Genes Mix With The Wrong Environment*," episode two of the series *Predict My Future, The Science Of Us* Razor films. Shown on TVO Jan 18, 2017. A transcript is available on line at http://tvo.org/transcript/119862x/video/documentaries/predict-my-future-the-science-of-us/ep-3-when-genes-mix-with-the-wrong-environment

dating back to 4000 BC, and he has never seen traces of the types of bone bruises that are typical of 'battered child syndrome.' In some modern societies, he estimates these marks would be found on the bones of more than one in 20 children who die between the ages of one and four.[260]

We need our families, but many modern governments encourage the break-up of families with welfare programs that trivialize parental authority. Children in a traditional family cannot claim full independence until they are able to make their own living. Until then, they are dependent on the older generation, and the older generation is responsible for them.

Like most systems based on family relationships, this one worked because it tied independence to a natural standard. When a child was able to take care of himself, he could leave the family home. Until then the family was responsible for him and was expected to teach him.

Government welfare programs free children from adult supervision, even if they need it. Some teen-agers are mature, of course, but these are not the ones who need welfare. Welfare caters to dysfunctional children who do not want to learn anything from adults or take responsibility for their own lives and who resent adult supervision. With welfare, they can avoid all education, supervision and control.

This increases the size of the welfare bureaucracy and therefore of the government as a whole. Because the welfare system trains people to rely on welfare it ensures more work for government bureaucracies but it teaches children to withdraw trust from families in favor of an impersonal and uncaring government.[261]

Day Care

Working mothers also create and justify a need for day care, which is another step toward eusociality. The first commercial day care centers opened a couple of centuries ago but there were few of them and they could not be considered important in their time. Day care in the Western world is now common enough to be significant.

Day care trains babies to fit into our culture and it frees women from family duties, but it is not a good way to raise human beings. Some animals bear their young in litters of up to a dozen at a time but human children are born one or two at a time and human mothers are not equipped to nurse more than two babies at once. Whether we assume that this is a result of evolution or perhaps planning by nature or by God, we know that human babies are not born to be raised in litters. They need individual care.

---

[260] Walker, Phillip, "Is the battered child syndrome a modern phenomenon?" *Proceedings of the Xth European Meeting of the Paleopathology Association.* See also "Battered Children," *Time* magazine, Aug 25/95, pg. 35.

[261] This problem is the theme of a selection of essays by Ivan Illich; Irving Kenneth Zola; John McKnight; Jonathan Caplan and Harley Shaiken, published as the book *Disabling Professions* by Marion Boyars Publishers Ltd., London, 1977.

One professional who studied day care and found it wanting is the late Dr. Burton White, a psychologist who has taught at Harvard, Brandeis and Tufts Universities. He was the founder of the Harvard Preschool Project and its director for the thirteen years it operated, the first director of the Brookline Early Education Project and senior consultant to Missouri's New Parents as Teachers project.

In a printed article he says: "I would not think of putting any child of my own into any substitute care program on a full-time basis, especially a center-based program."[262]

On his internet site about the problems of day care Dr. White says:

> The people that I know who have studied the development of children over the years number in the hundreds, because I've been around for a long time. I don't know two of them that applaud the notion of a transfer of the prime responsibility of child rearing over to any substitute. Most of the people I know do not like it. Very few of the people I know are willing to speak out in public the way I do. There's only two, there's Selma Fraiberg, and myself.[263]

If other psychologists are reluctant to criticize day care, that may be because many of them make their living from institutions that sponsor day care.

The effects of day care have been studied and the studies have been published but they are ignored by people who promote day care. Psychological experiments and studies conducted in the 1960s and 1970s found that babies are not happy when they are abandoned by their mothers. To solve this problem one American 'expert' who supports day care says that children will have less negative reactions to day care if they begin when they are about three months old, before they learn to recognize their mothers. If they don't know their mothers, they won't know they have been abandoned. If they have never been accepted, they can't be rejected.[264]

But that does not mean they are not harmed. Psychologist Selma Fraiberg says that babies who do not get attached to their mothers develop the condition she describes as "the disease of non-attachment" which prevents them from forming normal human bonds. With no

---

262 White, Dr. Burton "Viewpoint, should you stay home with your baby?" *Young Children* vol. 37 #1, Nov/81. The paragraph quoted is from the end of the article, on pg. 14; see also http://www.bostonglobe.com/metro/2013/10/17/burton-white-believed-children-should-avoid-day-care/DAH01Lkp5efC8pDKOWlKfI/story.html

263 I found this quote at cnet.unb.ca/orgs/prevention_cruelty/white.htm several years ago but my ISP can't find it now. In general, Dr. White's views are well known and have been quoted in literally hundreds of references on the internet. He has also published his views in several academic and popular books.

264 Blum, Marian, *The Day Care Dilemma*, D C Heath & Co, Lexington, Mass, 1983, pg. 18.

experience of human ties, she says, they may have no conscience and are often unable to observe or criticize their own behavior.[265]

Dr. White is not the only professional to have reservations about the value of day care. In 1989, Romania's National Authority for Child Protection invited a team of American experts — led by Charles A. Nelson III, Professor of Pediatrics and Neuroscience, and of Psychology, at Harvard Medical School; Nathan A. Fox, Distinguished Professor in the Department of Human Development and Quantitative Methodology at the University of Maryland and Charles H. Zeanah Jr., Professor of Psychiatry and Clinical Pediatrics and executive director of the Institute of Infant and Early Childhood Mental Health at Tulane University — to study the effects of institutional care on children.

Starting in 2000, the team chose 136 children aged six to 31 months, all free of defects, all of whom been sent to one of the six orphanages in Bucharest in their first weeks or months of life.

Divided into two groups at random, half were left in the institution and half sent to foster care. Because siblings were kept together, 53 families provided homes for 68 children. For comparison, the experimenters also recruited a third group of children who had never been institutionalized. The children were studied for ten years.

At 30, 40 and 52 months of age, the average IQ of the institutionalized children was in the low to mid 70s and that of the children in foster care about 10 points higher. children who had never been in an institution averaged 100, which is normal. Children who were moved to a foster home before they were two years old usually scored higher on IQ tests than those who were moved after two years in an institution.

By the age of 42 months, children in institutions were unable to form emotional attachments to other humans and Magnetic Resonance Imaging showed that they had less than normal volume of white and grey matter in their brains. The study also found that children who went to foster homes before they were 16 months old were able to develop normal language skills but those who were kept in institutions did not do as well. Children who spent any time in an institution had shorter telomeres than those who had not. Most doctors believe that cells with short telomeres age faster than cells with longer telomeres.[266]

Sweden has had nationally-subsidized day care since 1975 and 90 percent of all children 18 months to 5 years old are in daycare. A Swedish government investigation among Swedish 15-year-olds in 2006 showed that mental health declined from 1986 to 2002, based on self-reported symptoms such as anxiety, fright and alarm, and it declined faster than in 11 comparable European countries. The self-

[265] Fraiberg, Selma, *Every Child's Birthright*, Basic Books Inc. NY, 1977, pp. 43-51.
[266] Nelson Charles A, III, with Nathan A. Fox and Charles H. Zeanah Jr., "Anguish of the Abandoned Child," *Scientific American*, Apr 2013, pg. 62-67.

reports are confirmed by Folkhälsomyndigheten — the Public Health Agency of Sweden.

Suicide attempts among Swedish youth are also increasing. The Folkhälsomyndigheten report says it is not sure how to interpret these findings.

In 2005, a study found that the first generation of Swedish mothers who used the daycare system had an 'extremely high' rate of sick leave compared to other European countries. Anecdotal evidence suggests that this may be because mothers feel coerced, culturally and financially, to leave their one-year-olds in daycare.[267]

An article in Canada's *National Post* newspaper says that, working with data from Statistics Canada, a team of three researchers found "striking evidence" that exposure to Quebec's subsidized daycare program was associated with lower "non-cognitive skills," such as the ability to control impulsive behavior and maintain emotional stability. Boys show higher levels of hyperactivity and aggression and girls show declines in pro-social behavior such as donating and volunteering.[268]

A study of children five to eight years old who had spent most of their first years in a day care center at the University of North Carolina found that they were more likely to hit, kick, threaten and argue than children raised by mothers. Other studies have found that day care children are considerably less responsive to, and more aggressive to, adults than are children raised at home.[269]

As I write this bleeding hearts in Toronto, where I live, are lamenting the increase in 'bullying' in schools and on the internet, but they don't relate it to the prevalence of day care.

No surprise there, because — as mentioned earlier — modern psychologists find it more profitable to praise day care than to report the problems it can create. Perhaps they have not read the studies that report the problems.

Even if they had, these studies dealt only with the first generation of day care children who were tended, for the most part, by adults who were raised by their mothers. In future generations, when children in day care are tended by adults who were raised in day care, we can expect the effects to be more pronounced.

---

[267] http://www.imfcanada.org/archive/1107/swedish-daycare-international-example-or-cautionary-tale

[268] Kheiriddin, Rasha, "The Dangers Of Daycare," *National Post*, Toronto, Sept 24/15, pg. A14; see also "Daycare Skeptics" *The Globe and Mail*, Oct 22/1999.

Zinmeister, Karl, "Is It Good for the Children? Hard facts about day care," *Reader's Digest*, April 1989; and "Quebec's sacred cow has quality issues," *National Post*, Dec 6/04; and "Kids in daycare more likely to be overweight, study finds," *Toronto Star*, Nov 20/12, pg. A 1.

[269] Belsky, Jay, *The Effects of Infant Day Care Reconsidered*, US Department of Education, Office of Education Research and Improvement, Washington, 1988, pg. 30, 32, 33. See also Fredelle Maynard, *The Child Care Crisis*, Penguin Books Canada, Markham, 1985, pg. 119. See also Friscolanti, Michael, "Charges of liar greet key daycare study," *National Post*, May 7/01, pg. A1.

We know that after the people of Tasmania lost the secret of making fire they did not recover it until Abel Tasman 'discovered' the island in 1642. The techniques of motherhood that our species developed over thousands of years will probably survive for some time in some corners of the world but, while the people of Tasmania were willing to learn from Tasman, we in the 'developed' world do not learn from people we see as 'inferior.'

Within a few generations the experience of thousands of generations of mothers will be lost, and the nature of humanity will change.

On average people in the future will probably be more aggressive and less intelligent than we are, for two reasons. The prevalence of day care is one, and that effect will be compounded by environmental factors. Mercury poisoning, for example.

There is not much mercury in any one florescent tube or mini-florescent light bulb, but any large city has millions of them and they all burn out sometime. In theory, they could be disposed of safely but in practice they are just disposed of, and the mercury vapor flows through drains and sewers to the lakes and rivers that supply our drinking water.

In addition to the physical symptoms of Minamata disease — which include numbness in the hands and feet, loss of muscle strength and control, narrowing of the field of vision and damage to hearing and speech — mercury poisoning can cause aggressiveness, mood swings, and anti-social behavior.

We may also see less technology in our future, because women tend to be less mechanically-minded than men and, after we have used most of the world's resources, the technologists of the future may or may not find substitutes. On the other hand we might be safe to assume that — as in the utopian societies envisioned by Plato, St. Thomas More and Edward Bellamy — much of the physical work will be performed by slaves or otherwise forced labor. We already see this in the United States, where a mixed public and private prison industry holds the world's largest stock of prisoners and makes them work for commercial masters.

We don't hear much about the problems of day care in modern news media because most of the studies that found problems with day care date back to the 1950s, 60s and early 70s. Now the need for day care is part of conventional wisdom and modern research tends toward experiments that prove that, for selected slum children, very expensive care and training by psychologists offers some advantages over home care by inadequate parents. The studies don't report whether the 'inadequate' parents were raised by parents or in day care.

The experimental centers at which such studies are conducted bear no relationship to the average day care center, and the studies avoid any consideration of the damage that may be done to average children by the average level of day care. That's not surprising, given that much of the research is conducted by people who either make their living from day care or who plan to. They may not be consciously dishonest but they have an honest need to justify day care.

We do get occasional news that some day care operators may be deliberately dishonest or ignore regulations and safety concerns. Small day care centers that care for no more than five children outside the operator's family don't have to be licensed in Ontario but, after a child died in an un-licensed center in the Toronto suburb of Vaughan, provincial authorities who had ignored earlier complaints about the center finally inspected it.

They discovered that it had at least 35 children registered — seven times the legal limit for an un-licensed center — and they found dirty diapers in the kitchen, rotting food in the refrigerator and potentially deadly listeria bacteria in some of the food. Among the earlier complaints that were ignored was one that a three-and-a-half-year-old child, who was toilet-trained, was made to wear a diaper and belted into a car seat to watch TV, unsupervised, in the basement.[270]

Most modern governments enact and enforce regulations to govern the commercial preparations of food, but they don't seem to worry much about the commercial raising of children.

Some idealists pretend that day care workers are 'professionals' and that their care is educational and so forth, but that's an obvious shibboleth. A child at home watches his or her mother all day and goes with her to the store or whatever. Because the mother is devoted to the child she takes time to answer questions and, because the child sees the real world, he or she learns about the world.

Day care workers are not devoted to the children they tend and, with many children to look after, they can't take time to explain things or answer questions. When a child goes out — to a mall or to a store or to visit friends — with his or her mother, it's an adventure into the world and an introduction to reality. When a day care child goes out it's as one of three or four or even more kids in a group cart which the worker wheels through a mall or a park without stopping and without answering any questions, and the child learns nothing. The child belted into the car seat and left to watch television may be an extreme case, but it's obvious that a day care worker tending a half dozen or more children cannot give any of them individual attention.

It bothers me when I see a dog on a leash being dragged away from hydrants, trees and other things that it wants to stop and smell. It bothers me more when I see a child in a pram — or sometimes on a leash — being dragged away from something it wants to look at, or four or five children in a day-care stroller being wheeled through a shopping center with no time to look at anything.

Some psychologists favor day care and some oppose it, but psychologists who favor it enjoy the approval of the Mind Benders

[270] "Listeria found at Vaughan daycare," *Toronto Star* Sept 20/2013, pg. A1.

and, with funding for research, become better-known and their opinions are reported by the media. Psychologists who do not favor day are not supported, and their work is seldom recognized. We can also assume that this book will be criticized by supporters of day care.

The primary result — if not the intended function — of day care is to condition children to live within a eusocial society, but day care is also a system in its own right and it cooperates with other systems. As an addition to the cash economy, it offers a whole new industry for governments to license, supervise and tax. The licensing and the supervision are essentially meaningless but they create an additional function for governments, which in turn creates demand for more services within government and thus enlarges the whole system of the government.

Still more important effects of day care kick in after children grow older and leave the center. Children who are raised at home will adopt the values of their parents, and they may resist attempts by a propaganda machine to change them. Children raised by strangers will accept the values strangers give them.

Because they are not raised by their parents day care children do not learn to respect their parents or to look to them for leadership. If children resent being abandoned, they may resent or even hate their parents.

They may have to obey the adults and adolescents who work in the day care center but, because day care employees are paid help who don't have much time to interact with any one child, the children don't learn to look to them for approval.

In fact because most day care workers are low paid and some of them may resent their jobs but are not able to change them, children may hold them in contempt.

Children don't know about pay but they see their parents treat day care workers as inferiors and that the workers themselves feel impotent, and any young pack or herd animal soon learns who must be respected and who need not.

As noted earlier, children learn from peers and elders they see as high status and, as a general rule, they will not learn from adults they see as low status. With no high-status adult models to admire, day care children learn to identify with their peer group to a much greater extent than do children who are raised at home. If the high-status members of the peer group are bullies, bullying will be seen as admirable.

And whether they respect them or not, children need adults.

Even among animals, some migratory birds have to be taught to migrate and many carnivores have to be taught to hunt. If they have no elders to teach them, some birds will fly at the appropriate time but they may fly in circles or in the wrong direction.[271] Most carnivores

[271] Lorenz, Konrad, *Studies in Animal and Human Behavior*, translated by Robert Martin, Harvard University Press, 1970, vol. 1 pg. 80-81.

have a natural instinct to hunt, but unless they have elders to teach them, most will not learn to be efficient hunters.[272]

Human children need adults even more than other animals, because so much of our behavior is learned rather than instinctive. Instinct can't teach us how to make or use a bow and arrow or to plant yams or grain. Instinct does not teach us to make clay pots or stone axes, to weave baskets or to catch fish with a net or a hook. We learn all these behaviors, and we learn most of them from our parents and other adults in our families.

Somehow, we must learn the morals, mores, ethics, traditions and relationships that make our communities work. This has always been the responsibility of families, but the family's ability to teach depends on trust and respect for the wisdom of adults. If young people don't respect adults, they will not learn from them.

Breeding Like Insects

In 1932 novelist Aldous Huxley set out to portray the most dismally mechanistic and anti-human society that his fertile imagination could conceive. *Brave New World* started with the idea that humans could be 'hatched' in government hatcheries, rather than born, and that they could be raised by the state rather than by mothers.[273] We don't have government hatcheries yet, but more and more of our children are being raised in day care centers.

Even now, some women choose to have no children while others have many, and some women choose to be impregnated by artificial rather than natural means. In the foreseeable future, professional breeders could give birth 20 or more times each and, with modern techniques, they might produce twins or triplets every time for a total of 50 or more babies per breeder. This would be hard on the women, of course, but they could be well paid. With selective breeding and perhaps gene splicing, we might even produce a line of women specially adapted to high productivity. Even if they were not themselves ideal breeding stock, they could serve as host mothers for implanted fetuses.

Does this sound impossible? It is happening right now. The documentary film *House of Surrogates*, shown on the BBC4 network on Oct. 1, 2013, tells the story of a doctor in India who is building a huge clinic to house hundreds of poor women to live in while they carry babies for first-world customers. The doctor already has a smaller clinic that houses up to 100 surrogate mothers at a time. It has produced about 600 babies in the past ten years.[274]

---

[272] The wonderful book *Becoming a Tiger, How Baby Animals Learn to Live in the Wild*, by Susan McCarthy, published by HarperCollins, NY, 2004, describes the efforts of human care-givers to teach orphaned carnivores how to hunt.

[273] Huxley, Aldous, *Brave New World*, 1932, Bantam Classic edition, 27th printing, 1962. The government hatchery is described in the first chapter.

[274] "Birth of the 'baby factory,' *National Post*, Oct. 2/2013, pg. A1.

In Canada, one former resident of the co-operative building in which I live was a professed lesbian who had three babies, all by artificial insemination. The fathers are homosexuals who paid the mother to bear their children and they pay her to raise them.

That woman was an amateur but a Russian peasant woman of the 18th century shows us what the professional mothers of the future might achieve. Mrs. Feodor Vassilyev of Shuya, Russia, gave birth 27 times — bearing 16 pairs of twins, seven sets of triplets and four sets of quadruplets for a total of 69 children. Her husband's second wife produced six sets of twins and two sets of triplets.[275]

Even Mrs. Vassilyev could not produce as many offspring as a queen ant or bee, but that is not necessary. I suggest that human eusociality will be like that of termites rather than like ants or honeybees.

And we still might be able to match the insects, because individual women have millions of eggs each and a single man could produce enough sperm to fertilize all of them.

The 'government hatcheries' that Huxley imagined sound like science fiction, but they may not be all that far in the future. In April of 2017, the Children's Hospital of Philadelphia announced the success of its 'biobag' experiment in which lambs that had been prematurely removed from their mother were successfully maintained in an artificial womb. News reports suggested that the bag could support human babies that had been allowed to develop for 23 weeks in their mother's womb.[276]

When the artificial womb is perfected it will make it possible for baby factories to grow millions of siblings who, because they are siblings — possibly born or hatched at the same time and raised in the same environment — will have little individuality.

Termite colonies contain about as many males as females, but only one female lays eggs and most of the males are pseudergates who do not actually perform a male function. In modern human society we have countless gay men, some of whom behave more like females than like males and others who 'mate' only with other males. We also have technology that allows eggs from one female to be fertilized in vitro and implanted in a host mother. In essence, this is comparable to the insects' practice of keeping the queens' fertilized eggs in cells until they hatch.

As our unique form of eusociality develops we won't have exactly the same organization as ants and honeybees, but neither do termites; and mole rats; and meerkats do not have exactly the same eusocial organization as any insects.

---

[275]http://www.guinnessworldrecords.com/world-records/most-prolific-mother-ever

[276]https://www.theguardian.com/science/2017/apr/25/artificial-womb-for-premature-babies-successful-in-animal-trials-biobag

The Final Steps

So how will human eusocial society develop in the future? For one thing I suggest that the problems associated with day care now are just a start. If the care offered by day care workers who have been raised by mothers is inadequate — as it is — what can we expect of day care workers who have been raised in day care?

We can also predict that as women assume positions of power and men are relegated to menial jobs we may see the end of technical progress — which depends on people who look for alternatives to conventional solutions.

Women as a group are as intelligent as men but they are, on average, more average. On an IQ scale both sexes average the same score, but there are more men than women at both ends of the scale — with more than average and with less than average intelligence.[277]

Even among people of average intelligence, women are under more pressure than men to conform to an established norm. Since most of the pressure to conform comes from women, I don't see much chance that this will change in a society dominated by feminists; and if conformity is the norm, there is little room for new ideas. Our increasing insistence on formal schooling will increase this effect because, as noted earlier, formal schooling tends to reinforce the Einstellung effect.

In fact, it's not likely that a feminist-dominated society will maintain the technology we have now. Modern society encourages women to study and work in the sciences but most of our most important material progress is the work of mechanics and artisans, like James Watt, Henry Ford, Gaston Chevrolet, the Wright Brothers, Elisha Otis, John Deere, Clessie Cummins and countless others, and modern women are not encouraged or inclined to become either mechanics or artisans. We know that little boys and little girls tend to choose different toys and that among all mammalian species that have been studied, males and females develop different 'personalities.' If the leaders of society do not include people with much technical interest or ability, we can assume that technical progress will fail.

We see an example of questionable 'progress' in the enthusiasm for electric cars. Most people think they do not produce pollution but the facts are that about 65 percent of the electricity in the United States is produced by burning fossil fuels — 30 percent of the total from coal[278] — and that, because the energy is transformed several times — from heat to steam to torque in a turbine to electricity in a generator to storage in a battery — with a loss at every transformation, electric cars are less efficient and produce more pollution than comparable

277 https://www.psychologytoday.com/blog/sideways-view/201410/are-men-really-more-intelligent-women
https://en.wikipedia.org/wiki/Sex_differences_in_intelligence

278 https://www.eia.gov/tools/faqs/faq.php?id427&t3

conventional cars. Modern gas engines are about 30 percent efficient, which is about 10 percent less than the boiler that is the first step in a thermal power plant. Electric power producers tend to be shy about releasing efficiency numbers, but overall, they would do very well to achieve 20 percent efficiency from burning the fuel to where the power comes out of an outlet.

Even without electric cars we are using up the world's physical resources at an alarming rate and it's obvious that someday we will not have enough oil or any other kind of fuel to send bombers half-way around the world to destroy cities we have never seen and kill people we have never met. Personally, I don't see curtailing that activity as much of a problem, but there are some uses for oil and other resources that I think we will miss when they run out.

The overall nature of humanity is also changing, by evolution and as a side-effect of day care. Author and philosopher Jared Diamond suggests that among 'civilized' people evolutionary pressure tends to select for resistance to disease rather than for intelligence[279] and — as we have seen —studies of children raised in day care find that they are less intelligent on average than children raised by their mothers. If we assume that the most intelligent women are likely to have the most resources and be most able to commit their children to day care, we can predict a general decline in human intelligence.

We also have the problem that planners — both male and female — are gaining more and more power in modern society. I expect this trend to continue as feminists gain power, because women are better than men at accepting and following plans, especially plans made by others. I have no doubt that women planners are just as likely as men to make plans that suit their own purposes and to ignore or even conceal their effect on society as a whole.

The rise of feminism is one of the final steps toward full eusociality. As noted before, eusocial societies are dominated by females and, I suggest, domination by females is required for full eusociality.

The reason is simple and basic. Females cooperate; males compete. This is inherent in the roles of the two sexes — females cooperate because that's the best way to raise young and males compete because they are the wild card in evolution and, by competing, they establish their fitness to breed. If some males compete for the right to breed, all males must compete if they wish to breed. All males have an inclination to compete because they are descended from successful competitors.

Granted that many women in the modern world appear to be intensely competitive, I suggest that this is an illusion. In fact they are cooperating with the doctrine that women are just like men and can, and should, compete with men.

A noted earlier, it's easier to ride a mare than a stallion and easier to herd cows than bulls.

[279] Diamond, Jared, *Guns, Germs and Steel*, WW Norton and Co, NY, 1999, pg. 20-21

Recap

So, where are we now? I guess it's time to recap my argument.

Most people think humans are the dominant species on Earth but ants — which are more than 100 times as old as we are — outnumber us by millions to one and have a more highly developed culture. Termites, which are also eusocial, have been around even longer than ants but are not quite as successful. Both of these, and several other species of insects, thrips, arachnids, shrimp and even a few mammals, are eusocial.

Biologists now use the term to describe animals living at the highest level of socialization, with two or more generations in nests or colonies that they will defend against intruders and that will not be abandoned when the first occupants die, with a caste system, division of labor and, in some cases, only one or only a few breeders. In some eusocial colonies only the queens breed and in most the young are cared for by workers, rather than by their mothers.

Many biologists agree that we are eusocial too, if not quite as completely so — yet — as the insects. The accepted standard is that a species is considered to be eusocial when it has two or more generations living together in nests or colonies that they will defend against intruders and that will not be abandoned when the first occupants die; plus a caste system, division of labor and only one or only a few breeders.

Some suggest that eusociality develops because the members of a colony are close relatives but some eusocial species — even some varieties of ants — will allow outsiders to join their colony. Some suggest that eusociality develops when a group establishes a common home but some varieties of ants are nomadic, with no permanent home.

I suggest that species can be described as eusocial if an individual cannot live without help from others of its species. Some people think they could survive in the wild but even the best woodsman needs a knife or an axe, and it's not easy to find the right stone to make one. Even in stone age cultures flint was mined on an industrial scale and obsidian was traded over thousands of miles.

We're not completely eusocial yet because some of us can still think independent thoughts, but some very big and wealthy industries are trying to change that. I hope they don't succeed because while I don't knock the ideal of eusociality, I think we have gone far enough.

Our first step toward eusociality was to form a herd, and that was a simple process. In 1986, computer programmer Craig Reynolds wrote a program in which three rules make a flock of virtual 'boids' behave like a flock of birds or a school of fish. We can see how flocking, schooling and herding behavior would develop because animals in flocks, schools and herds are more likely to survive than they would as individuals.

Even in herds, individuals don't survive for very long and when we speak of 'survival of the fittest' we refer to survival of the species, not of the individual. Biologist E.O. Wilson estimates that the average survival of a species of mammal is about a half million years, while ants have been around for at least 100 million years, and termites for perhaps 150 million. Humanity is about one million years old, and it's safe to assume that we survive because we cooperate. If we don't survive much longer — which is a possibility — it will be because we do not cooperate well enough.

In a short run of the classroom game of Prisoners' Dilemma, the best policy is to betray your opponent but a computer tournament proved that, in the long run, a strategy of cooperation is unbeatable.

I suggest that all human behavior can be divided into two very general categories, survival positive behavior and survival negative behavior. In both cases the survival that counts is the survival of the group, herd or species rather than of the individual.

The great apes are stronger and more intelligent than baboons and they are protected by law, but they are in danger of extinction while the numbers of baboons, which are more cooperative, are not.

Herds don't need to organize cooperation but hunting packs do, and early humans were pack hunters. While most animals learn by imitation, we have the advantage of speech and, for the past several thousand years, writing.

Speech has played a major role in human development. Most animals learn by imitation, but with speech we can describe things that we don't have immediately available and behavior that we can't demonstrate. Reading and writing extend this power to the extent that we can learn from people we never meet and even some that died before we were born.

Fire is also important. Partly because we learned to use it long before we learned to make it, fire encouraged our ancient ancestors to settle or, if nomadic, to move from one established campsite to another. We know that in historic times the people of Tasmania knew how to use fire but not how to make it.

Conventional wisdom holds that our ancestors came down from trees to live on a dry savanna, but we have physical evidence that we are not well adapted to life on a savanna. Sir Alister Hardy's aquatic ape hypothesis, physical evidence and our modern preferences all suggest that we are more at home on the shores of lakes or the sea.

Some archaeologists argue that we have no evidence of early man and/or settlements on coastlines, but they ignore evidence that sea levels have risen hundreds of feet in human times and that the coastal areas that may have been home to early men are now tens or hundreds of miles offshore and deep under water. One estimate suggests that the total area of land submerged in the last 20,000 years is about equal to the area of Africa.

Conventional wisdom also postulates an agricultural revolution in which ancient people invented farming but this is based on ideas developed by 'scientists' of an earlier age, when most advanced education was essentially the elaboration of conventional wisdom. Modern anthropologists know that the development of grain farming did not improve living conditions for the mass of humanity.

The 'proof' that farming preceded settlement is based on genetic evidence of the evolution of 'domestic' grains and studies of ancient ruins — which show that the grains are much older than the ruins of any ancient city — but I argue that the grains we see as 'domestic' are more likely to have evolved naturally and that the ruins of our first villages, towns and cities are probably — like some ancient ruins that we know of — lost under the sea.

Early pre-historians thought the lives of nomadic hunters and gatherers were hard but modern anthropologists who have lived with them for months at a time report that their lives are easy, their diets generally better and their food supply more dependable than those of farmers. Some hunting and gathering people collected grains, but they did not have to farm them.

Both the trees-to-savanna convention and the theory of an agricultural revolution are generally accepted, but evidence suggests that both are invalid. This is important because if you start wrong, it is not likely that you will end right.

Even if you know which direction the bow of a ship is pointing you can't tell where the ship is going unless you know where it started and/or have full information on the winds and currents that affect it and how much the captain knows about them. Even if you just know where it started you can infer a destination if it is on the same ocean as the port of departure, because extension of the line from where it started to where it is now might show where it will probably wind up.

The late author and philosopher Jane Jacobs argued that people settled in villages before they began farming. Many professional archaeologists don't buy her argument but I do, and I extend it to cover the development of trade and robbery, armies and the conquest and occupation of craft villages.

Jacobs rationalized a scenario based on reports of the excavation of the site of Catal Huyuk, in Anatolia, which seems to have been a center for stone knappers who apparently imported their raw materials from elsewhere in Anatolia and exported finished products to Europe and elsewhere. I infer that an artisan might choose to settle close to a source of raw materials, or to materials he or she might have stockpiled and left to dry or season, that others might choose to join him or her and that the camp might develop into a craft village. Some technology that we see as 'primitive' is, in fact, quite sophisticated.

The establishment of a craft village would have driven the development of trade, trade would have spawned robbery, and robbery would have driven the development of trade caravans which would have driven the evolution of large robber bands. If robber bands had joined forces they would have become an army — which would include strangers as well as friends and would have a different ethos from a hunting or raiding band — and an army might conquer and occupy a craft village. The conquest and occupation of the village established the pattern of predators who have lived on the work of productive people through most of human history.

If grain that appears to be domesticated evolved around a village the bandits who occupied the village might have enslaved the villagers and forced them to farm the surrounding fields. Through most of history big farms have been worked by sharecroppers, serfs or slaves.

Even though they worked hard and produced more food than they could eat most of the farmers of history have not been as well fed or as healthy and did not live as long as hunter/gatherers.

Hunting and gathering bands and craft villages were associations of friends and relatives, but the occupied village was like a tropical reef in which some of the residents preyed on their neighbors. This pattern still exists in modern towns and cities.

Philip Zimbardo's Stanford Prison Experiment demonstrated that normal people may become bullies and sadists when they hold positions of power. We can assume that the bandits who conquered and occupied the craft village did the same.

With the bandits in charge the village developed a caste system with the bandits on top, villagers who could be of direct use to the bandits in the middle and others — some of whom may have been enslaved, on the bottom.

The occupation of the village also produced the cash economy, which has been a blessing and a curse ever since. It's a blessing because it enables thousands of people who will never meet and who have no common interest to collaborate in the production of artifacts as complex as jet airliners and it's a curse because it enables people to amass huge fortunes without doing anything useful to justify their existence while, at the same time, it keeps child laborers at work around the world. It's an unfortunate fact that the possession or control of a lot of money confers prestige no matter how the money is gained.

Biologist Marston Bates suggests that life forms can be sorted into three groups that he calls producers, consumers and decomposers.

Producers, mostly plants, produce food — mostly by photosynthesis. Consumers, mostly animals, eat either producers or the fruit or seeds they produce or, if they are carnivores, other consumers. Decomposers, mostly bacteria, decompose dead producers and consumers.

In the occupied village I identify two distinct groups of humans that I call makers and takers and two sub-groups — traders and agents — members of which align themselves with one or another of the main groups. These designations refer to occupations rather than to individuals and it is at least theoretically possible for one person to be a member of all four groups at different times.

The modern world could not function without some people that I would categorize as takers and I accept that, but because takers have an easier life and are often paid more than makers, most people would rather be takers than makers. Our modern economy can support so many takers only because we prey on makers, many of them children, in other countries, and we are willing to tolerate the poverty and misery of millions of would-be makers in the 'developed' world who are unemployed because people in less-developed countries work cheaper.

Members of a herd or a eusocial society mostly think alike because they are close relatives and live together. Artisans in a craft village probably hold many of the same ideas but, because they have different skills, they don't all think alike. The take-over and occupation of the village introduced a new herd — the army — with totally different skills and attitudes, and forced a major change in the nature of human society.

In modern times most of us get our conventional wisdom from schools. We all get the same basic schooling, but future makers go on to 'training' and future takers proceed to scholarly 'learning' — much of which has little or no practical use but which distinguishes them from the masses and qualifies them to become takers.

Schools teach their students to adhere to schedules, to believe what they are told by 'authorities' and to depend on authorities to provide ideas and solutions rather than look for them ourselves. Schools also produce the phenomenon that psychologists call the 'Einstellung effect' which leads people to apply learned solutions to all problems, even when a simpler solution is obvious. Philosopher Ivan Illich suggested that schools also teach us the need to be taught rather than to work things out for ourselves.

In fact the training of a typical hunter/gatherer is far more extensive and useful than the education of most modern people and, while a hunter/gatherer can learn enough to integrate into modern society in less than a year, it would take a typical 'educated' modern person five or ten years to learn enough to function as a hunter/gatherer.

Our culture of schools and rulers also distorts our social order and our perception of the contributions other people make to our welfare. In fact the job of a mechanic may be more demanding than that of a doctor and a bus or truck driver works harder and has more personal responsibility than an airline pilot. We are less affected by the months of vacation taken by CEOs, senators or members of Parliament or Congress than we are by a two-week shut-down of garbage collection or sewage treatment.

Advanced education has the advantage that it keeps people neutralized in school through most of the years in which they might be vulnerable to new ideas. In many countries it also builds up a load of debt that helps to discourage any idea of rebellion. Because many people get used to debt it becomes part of their lives, and millions of us keep savings accounts that typically pay three percent interest while at the same time we support a credit card debt that often costs 18 percent interest.

We need advanced education to support some developments of modern technology but the people who developed most of that technology did not learn it in school and some of our most prolific inventors were apprenticed rather than 'educated.'

Education also encourages us to make plans rather than to rely on swarm intelligence or allow situations to develop naturally. Short-term plans are valuable to hunters but the long-term plans made by rulers are often failures, because we can't see far enough ahead to avoid pitfalls. We don't recognize the failure of many plans because it's often easier to fudge results than to admit failures.

Whether it's the sole cause or not, I argue that reliance on plans is one of the reasons that while some 'simple' hunting and gathering societies have lasted tens of thousands of years, no 'civilized' society has lasted more than a few hundred years. the usual cause of collapse of a hunting and gathering society is conquest by the army of a 'civilized' society, but the usual cause of collapse of a 'civilized' society is internal failure.

Author Jane Jacobs outlined several examples of disastrous redevelopment plans in American cities. A ban on alcoholic drinks in the United States played a key role in the development of organized crime and a plan to reduce drug use has nurtured crime in the United States, created and supported wealthy and powerful anti-social gangs in Mexico and given the United States the world's biggest and most oppressive prison system.

The 'ideal' societies envisioned by Plato, St. Thomas More and Edward Bellamy were all dictatorships or bureaucracies populated by people who didn't have much personal freedom and supported largely by involuntary labor.

In contrast, Prof. Tadeo Umaseo attributes the modernization of Japan to the process that he describes as 'guided drift.' Rather than plan ahead, he says, the Japanese tendency is to watch developments as they occur and adjust policy to accommodate them.

Robbers conquered the craft village but priests helped them rule it and priests have supported most rulers since. Religion can be a unifying force and priests who support the rulers' religions generally do very well for themselves. At some points in history they have been the most powerful, and the greediest, forces in the land.

The power of priests has been supplanted in some countries by the propaganda and advertising industries, which have developed an arsenal of psychological tricks to distort people's beliefs and perceptions and to control their behavior.

French scholar Jacques Ellul distinguished two types of propaganda, with two subdivisions of each. 'Agitative' propaganda tries to convince us to change something and 'integrative' propaganda tries to convince us that there is no need for change. Either type can be either 'political' propaganda, which is disseminated by a political entity with a specific program or 'sociological' propaganda, which is disseminated by all of us — often unintentionally.

I recognize a third type, 'collateral' propaganda, which I consider to be the most pervasive and most dangerous of all. Like collateral damage in a war, collateral propaganda misses or overshoots its intended target to affect people it is not aimed at.

An advertisement for an expensive car, for example, is aimed at people who can afford to buy expensive cars but is seen by many people who can't afford any car and it may depress them or make them angry. A policeman's or a politician's warning about the threat of terrorism or crime may be an attempt to get more funding but it may also encourage terrorists and criminals, or frighten people who are afraid of them.

One example of collateral propaganda is the phenomenon I call 'Barnum's law,' which states that "there is no such thing as bad publicity." Because of Barnum's law, the American 'War on Drugs' actually promotes the use of drugs in the United States while, as a side effect, it has created and supports expensive and powerful police forces and the world's biggest prison system in the United States and powerful criminal gangs in Mexico, Colombia and other countries. In Canada an attempt to limit domestic violence, and a separate program to reduce the use of handguns among criminals, have exacerbated both problems.

Because propagandists and advertisers live in a maelstrom of propaganda and advertising, they may lose sight of the real world and inflate their own claims to make them even more impressive.

Some people think their education protects them from propaganda, but in fact it makes them more vulnerable. 'Belief' is created and supported by five factors that I call primacy, authority, imitation, comfort and repetition. Advertisers and propagandists can and do use all of them to implant their messages and/or beliefs in our minds.

Many of us are controlled by the phenomenon of 'groupthink' in which the members of a group echo the thoughts and beliefs of their leader. This has been blamed for many of the wars and other disasters of world history.

Advertising is a close relative of propaganda and, like propaganda, it includes subliminal techniques that can implant messages that we may not be conscious of hearing or seeing. Many people believe that subliminal techniques are illegal and/or that they don't work, but in fact they are legal and they work very well.

Subliminal advertising was caught in action during the summer of 2000 when news programs revealed the hidden message in a commercial supporting presidential candidate George W Bush. When the commercial was slowed down, we saw the word "rats" drifting across a picture of Bush's opponent, Al Gore. Bush's lead in the race decreased after the subliminal message was revealed, but he still won.

Because they are our most powerful mediums of communication movies and TV shows set the norm for our society and it should be no surprise that the violence and crime in movies and on TV is reflected in real life. In fiction the policemen who shoot first are always right and their victims are always guilty, but it doesn't work that way in real life.

Most individual advertisers and propagandists promote their own interests but, as a group, the Mind Benders deliberately warp our perceptions of the world and the development of our society. The metasystem is dangerous because the media that many of us trust depends on advertising to operate, and the people who control the advertising control the media. Because that control is second-hand, advertisers need feel no responsibility for the ideas they support. In fact, I suggest, most propagandists and advertisers are victims of their own propaganda.

J.K. Galbraith, who knew Ronald Reagan when he was an actor, wrote that Reagan was president of an actors' union and actively liberal before his acting career failed and he was hired to deliver a series of right-wing speeches.[280]

In the Western world, most of the beliefs the Mind Benders promote appear to be harmless or at least justified, but many are dangerous. Many of the kingdoms and empires of history were overthrown because the takers flaunted their wealth, and the makers who produced the wealth they were not allowed to share lost patience. In the modern world, movies, television and print media all show us a glittering world of privilege that we cannot share, and in some cases the image is so exaggerated that not even the most parasitic of takers can attain it.

The development of chain newspapers and radio and TV programs also promotes the power of propaganda because all members of a chain take their opinions and information from one source, and there may not be a serious attempt to control or counter false information. Media chains, the advertisers that support them, the journalists they employ and the politicians and activists they support form a metasystem that is essentially autonomous and that has no interest in the welfare of humanity.

We have seen the deliberate destruction of human families for the benefit of commercial interests and the replacement of the parental instruction that is the birthright of all humans with programmed beliefs and values that serve the advertisers best.

---

[280] Galbraith, John Kenneth, *Name-Dropping, from F.D.R. on.* Houghton Mifflin, Boston, 1999, pg. 3

Many people are convinced that 'the good life' is based on geegaws and video games while hundreds of thousands of our own countrymen go hungry and have no home to live in. We waste billions of dollars on make-believe exploration of other stars and planets we will never visit, and people who have never explored their home town dream of exploring outer space.

Mind benders tell us that youth has always been rebellious, but they try to hide the fact that they are leading the rebellion and that they are leading it in directions that are good for them rather than for the young 'rebels' they lead.

In recent years the Mind benders have moved from alienation of the younger generation to alienation of women. This is partially justified by fear of over-population, but the justification is invalid because we know from observation that people in poverty tend to breed more than people who are prosperous — partly because they can't expect all their children to survive and partly because they will need children to support them in their old age.

The women's movement discourages women from productive work at home and encourages them to work in the commercial economy where they depress wages and increase costs, often while performing work that has little real value.

But the women's movement is a vital step in the development of complete eusociality because all known eusocial communities are dominated by females. Male ants and honeybees live only to breed, which they do only once, and most male termites — like some modern men — are essentially feminized. This should be no surprise because, as the computer tournament of the Prisoners' Dilemma game demonstrates, cooperation always beats aggression in the long run and herding and cooperation instincts are stronger in women than in men.

I suggest that the 'battle of the sexes' began with the conquest and occupation of the village. Women in hunting and gathering communities are free because they don't have to rely on men, but if women in an occupied village don't have a man to protect them, they may be enslaved.

The modern women's movement victimizes children by breaking up the families they need to support them. This is a tragedy because a step-child is about forty times more likely to be abused than a child living with both natural parents and, in some areas of the world, children who live with two biological parents are a minority. Some children who are tortured, starved or beaten to death by step-parents make the news, but most don't.

Paleopathologist Phillip Walker has examined the bones of more than 5,000 children from preindustrial cultures dating back to 4000 BC and, he says, He has never seen the bone bruises that are evidence of 'battered child syndrome.' He estimates that, in some modern societies, more than one in 20 children who die between the ages of one and four are victims of step-parental violence.

When a male lion takes over an existing pride, he kills all the cubs of the deposed male. We pretend that humans are different.

The women's movement also supports the development of day care, which leaves women free to work in the commercial rather than a home economy, and subjects children to programming by minimum-wage workers who have no emotional attachment to them. Studies in several countries have shown that children raised by institutions are, on average, less intelligent and more aggressive than children raised by parents, but many 'modern' mothers are still enthusiastic supporters of day care.

One continuing study has found that children are less affected by stress when they are with their families than when they are away from them.

Care of the young by workers rather than their mothers or other close relatives is a new development among humans, but the norm in eusocial societies.

And day care is just a start. We already have professional surrogate mothers who carry and bear children conceived in vitro, from the eggs of another woman and a man that neither woman has met. We assume that few surrogates bear many more children than natural mothers, but in theory at least, they can.

The *Brave New World* that author Aldous Huxley envisioned in his dystopian book bred children in artificial wombs, and, in our world, an artificial womb has been developed and is being tested.

The combination of feminism, daycare, surrogate motherhood and an artificial womb could make us completely eusocial. This may ensure the survival of our species, but that does not mean that we would recognize the people of the future as human.

Evidence suggests that most people in the future will be more aggressive and less intelligent than we are, they will have less technology than we have, and some form of slavery will be re-introduced.

We have already seen that psychopaths who do not have much humanity can gain an advantage in our system-ridden society, and we must assume that the super-psychopaths of the future — who may be conceived in glass dishes, nurtured in artificial wombs and raised in day care centers — will eventually rule the Earth. I consider this more or less inevitable.

We see the trend already in studies of the first generation of day care children, most of whom were tended by workers who were raised by their own mothers. In future generations, when children in day care are tended by workers who were raised in day care, we can expect the effects of day care to be more pronounced.

Whatever the reason, we also know that there is a potentially disastrous drop in sperm count and motility among men in the

Western world. This may be an environmental rather than a cultural problem, but it is occurring among people who have considerable control over their environment.

Assuming that the Western World survives I predict that people in the future will have less technology than we are used to. On the other hand we might be safe to assume that — as in the Utopian societies envisioned by Plato, St. Thomas More and Edward Bellamy — most physical work will be performed by slaves or otherwise forced labor, such as the child workers of the third world or inmates of the American prison industry.

Plato pretended that the leader of his ideal society would be a 'philosopher king' and that his cohorts would also be philosophers, but if we look at what he actually wrote rather than how his work is presented by his groupies, we can see that his dream is twisted, by modern standards. Thomas More and Edward Bellamy don't say much about the leaders of their idealized societies but we have no reason to suspect that they would be democratically elected or subject to the same restraints and conditions as the population in general.

Even now, we know, a small group of elite live lives totally different from the rest of us and we can assume that this will continue. Through most of history we have seen kingdoms and empires overturned by revolution when inequality and repression became too blatantly obvious but the elite of the modern world — if not in this iteration then in the next, or one after that — have enough weapons and private armies to prevent revolution. If the guards of the Bastille had been equipped with flame throwers, France might still be ruled by a Sun King. In the future, it may be again.

I do not suggest that governments of the future will use flame throwers on the masses, because they won't need to. The Mind Benders rule by bending minds rather than by breaking bodies, and the modern world has developed some very sophisticated tools for bending minds. If that's not frightening enough, consider that these tools are non-directional and they affect the people who use them just as much as they affect the people they are used on.

Through most of the era we see as history, most of humanity has been ruled by the metasystems that controlled our rulers rather than by the individuals we saw as rulers. The Baron could rule the occupied village only with the support of his army and priests, the army consisted of officers who needed the support of their soldiers and the priests needed the support of their worshippers, and so forth. In every case the titular head of a system was in fact the servant of a system and individual systems survived because they served a metasystem.

Metasystems survive and prosper if they serve their own needs and those of people who support the metasystem.

Wrapping It Up

I'm an old man now and I don't expect to see the end of 'civilization as we know it,' but I suspect that some people now alive, or who will be born in the next few years, will. I pity them.

Not the kind of future we hope for, is it? Humanity with the habits — and perhaps the brains — of ants. Okay — here's something nicer to wind up with.

Ants are descended from wasps, which have probably been eusocial for more than 100 million years.

But some varieties of wasps — called mud daubers — build individual nests and raise their own young in them. In most cases it's females that build the nests and feed the young, but some males have been seen building nests, protecting them and bringing spiders — food for wasp larvae — to them.

There are even a few species of ants that live as individuals — the venomous *Myrmecia pilosula*, commonly known as the hopper ant, jumper ant or jumping jack, in Australia and *Harpegnathos saltator* in India. The insect many people in Kentucky call the 'velvet ant' also lives as an individual, but entomologists tell us that it is actually a "wingless wasp." There may be other individualist ants, but they are hard to find and their numbers are insignificant compared with the eusocial species. Nonetheless, they do exist and there may be — probably are — more than we know of.

And while most of humanity follows the crowd, some people live as individuals — doing their own thing their own way — and some do very well at it.

So while most of humanity is eusocial, that doesn't mean you have to be part of the mindless mob. There will always be room for a few individuals, and you could be one of them.

Made in the USA
Monee, IL
11 November 2026

3785886e-a469-4f1a-890d-0152ad0d205fR01